J. G. Wood

Common British Insects

Selected from the Typical Beetles, Moths, and Butterflies of Great Britain

J. G. Wood

Common British Insects
Selected from the Typical Beetles, Moths, and Butterflies of Great Britain

ISBN/EAN: 9783337139803

Printed in Europe, USA, Canada, Australia, Japan

Cover: Foto ©berggeist007 / pixelio.de

More available books at **www.hansebooks.com**

COMMON BRITISH INSECTS

SELECTED FROM THE TYPICAL

BEETLES, MOTHS, AND BUTTERFLIES
OF GREAT BRITAIN

BY THE

REV. J. G. WOOD, M.A. &c.

AUTHOR OF
'HOMES WITHOUT HANDS' 'BIBLE ANIMALS' 'COMMON OBJECTS
OF THE SEA-SHORE AND COUNTRY' ETC.

WITH ONE HUNDRED AND THIRTY FIGURES BY E. A. SMITH
ENGRAVED BY G. PEARSON

LONDON
LONGMANS, GREEN, AND CO.
1882

PREFACE.

IN A WORK entitled 'Insects at Home' I have described and figured the most conspicuous examples of every order of British Insects.

That book is necessarily a work of some dimensions, and occupies nearly seven hundred pages. It has been suggested that as more interest is generally taken in the Beetles, the Butterflies, and Moths, than in the other orders, it would be as well to publish an abridged account of those two orders. This has been done, and the reader will find embodied in the present work some of the more important discoveries which have been made since 'Insects at Home' was originally published.

J. G. WOOD.

CONTENTS

LEPIDOPTERA.

COMMON BRITISH INSECTS.

CHAPTER I.

INTRODUCTION.

THERE is scarcely a branch of science, however interesting it may be, which does not at first repel the intending student by the array of strange words with which the treasures of knowledge are surrounded. This is especially the case in Botany and Zoology, which contain, in addition to the usual technical language, vast numbers of names belonging to various plants or animals, each name consisting of two words, one denoting the genus and the other the species.

In the following pages I intend to describe, as far as possible within so limited a space, the butterflies moths, and beetles of Great Britain, and, though giving the needful scientific information, to use few technical terms, and always to explain those which of necessity must be employed.

OUR first business is evidently, when treating of these insects, to define precisely what an insect is,

This seems to be a simple matter enough ; but it really is not so, the question being one which has occupied systematic zoologists for many years, and which is even now rather a dubious one in several cases. The word insect is, as a rule, employed very loosely by those who have not studied the subject. Spiders, for example, are generally called insects, and so are woodlice, centipedes, and a variety of other creatures which have really no right whatever to the title. We will therefore see what an insect really is.

Insects are technically described as being '*articulated animals, breathing by tracheæ, divided into three distinct portions—viz. the head, the thorax, and the abdomen—passing through a series of transformations, and having in the perfect or "winged" state six articulated legs and two antennæ.*'

We will now take this description and examine it in detail. The articulated animals are formed on a totally different plan from the vertebrates, molluscs, radiata, or other divisions of the animal kingdom. Their bodies are formed of a series of more or less flattened rings, within which are contained all the muscles and vital apparatus. It will be seen that a vast number of animals come within this definition, which includes not only the insects, but the Crustacea, such as the crabs, lobsters, shrimps, woodlice, and others ; the Arachnida, such as the spiders, scorpions, and mites ; the Myriapoda, such as the centipedes and millipedes ; and the Annelida, of which the common worm is a familiar example. It is necessary, therefore, to find some mode of distinguishing the insects from all the other articulates, and, after much trouble, systematic naturalists have agreed upon the short formula which has already been given.

It is there stated that insects breathe through 'tracheæ.' Now tracheæ are tubes composed of thin membranes, kept open by a fine but stiff wiry thread, which is twisted spirally throughout the whole course of the tube, just as a modern flexible gas-tube is kept open by a spiral wire, no matter how it may be twisted or bent. This is absolutely necessary in insects, for the tracheæ are not confined to a single portion of the body, like the lungs of men or the gills of fish, but permeate the entire insect, passing through all the limbs, and even reaching to the claws which terminate the feet. Any of my readers who wish to see the extraordinary manner in which the breathing apparatus is disposed over the whole body should look at the plates of Strauss Durckheim's wonderful monograph on the common cockchafer, a work to praise which would be simply impertinent.

I strongly advise all my readers to examine these marvellous structures for themselves. There is not the least difficulty in finding them, for the real difficulty is to dissect any part of the body without finding them. The largest of these tubes are those which run along the sides of the insect, and are connected with the oval openings along the sides, which are possessed by every insect. These openings are called spiracles, from the Latin word *spiro*, because through them the insect breathes. Any insect or caterpillar will furnish the tracheæ, but the larger the better. They should be severed from the body by a pair of fine scissors, then taken out with a pair of forceps, and laid on a glass slide. I have now before me a preparation of the tracheæ of a silkworm which I made twenty-two years ago, and it is not the least damaged by keeping.

These tracheæ afford a most important characteristic of the insects, inasmuch as the Crustacea do not possess them at all, and the Arachnida generally, though not always, breathe by means of internal air-sacs.

Next, the creature must be divided into three distinct portions. This is the signification of the title Insect, which is derived from two Latin words, signifying cut-into, while the familiar Greek name of Entoma (from which the word entomology is formed) has precisely the same signification. This is, perhaps, the most important of all the characteristics, as in the Crustacea and Arachnida the head is merged into the thorax, so that they are divided into two portions instead of three; while in the Myriapoda and Annelida there is no distinct thorax, and sometimes scarcely a distinct head.

Next we come to the transformations which insects have to undergo before they reach their perfect or adult state. All animals really undergo a course of transformation, but in the insect they take four very distinct forms; namely, the Egg, the Larva (i.e. caterpillar or grub), the Pupa (or chrysalis), and the imago, or perfect insect. Any of my readers who have bred silkworms will be practically acquainted with this fact, and will also know that the larva changes its skin, or moults, several times before it assumes the pupal form. The reason for this casting of skin is evident. The larva, like the perfect insect, is made of a series of flattened rings, or rather, of a double series of half rings, connected along the sides by an elastic membrane, so as to permit the creature to breathe and eat.

Now, the upper and lower portions of these rings are comparatively inelastic, and cannot themselves expand, though they can be opened wider at the sides in proportion to the interior expansion of the body. Meanwhile, the larva continues busily its sole business, that of eating, and increases rapidly in size, so that, within a certain time, its skin is stretched to the utmost, and can expand no more. Still the larva continues to increase, though its tight integuments cause it so much uneasiness that it ceases to eat, and at last the overstretched skin bursts, and the larva emerges, clad with a new skin, which has been forming under the old one. As soon as it is free, it takes a number of deep respirations, and in half an hour, or thereabouts, is so much larger than its cast skin, that to put it back again would be impossible. This process is repeated until the larva is about to assume the third or pupal state.

In consequence of this mode of development, the whole of the growth is completed during the larval state, and, however long an insect may live, it never grows after it has attained its perfect form ; and, though it is common enough to find insects though of the same species yet of very different sizes, the largest have not grown since their last change, nor will the small specimens ever attain the dimensions of their larger relatives. In a measure, the same rule prevails among mankind, and, though some may be giants and others dwarfs, the dwarf will never become a giant, nor has the giant ever been a dwarf, and, different as are their sizes, both ceased to grow when they attained the age of manhood.

The modes of passing through the successive changes of form are exceedingly variable in the different orders of insects, and are always most interesting to careful observers. I shall not mention them in this place, but shall give the descriptions of the metamorphoses together with the history of the different species.

Next in order comes the statement that all true insects have six legs when they have attained the perfect form, or Imago. The reader will see that this definition at once excludes all other Annulata. The Crustacea, for example, have a considerable number of legs, and the Arachnida are eight-legged, while the Myriapoda are, as their name infers, many-legged, and the Annelida have no legs at all. It is true that in some insects there only appear to be four legs, but, in these cases, the apparently missing organs may be discovered on careful examination, much reduced in size, but still present.

A similar observation may be made with regard to the antennæ, or, as they are popularly called, 'horns,' or feelers.' The word antenna is a Latin one, signifying the yard-arm of a ship, and has been appropriately given to these organs. In most cases the antennæ give great character to the aspect of an insect. In some of the Beetles, for example, they are slender, and each joint is so lengthened that the antennæ are five times as long as the body. In others they are comparatively short, sometimes deeply toothed like combs, sometimes terminated with a round club, sometimes with the ends developed into a beautiful fan-like apparatus, and sometimes looking like a number of coins joined together by a string running through their centre. The knob-tipped antennæ of the butterflies are an unfailing characteristic whereby these insects can be distinguished from the moths, with their sharp-tipped antennæ ; while in the latter group of insects, the antennæ of the male are often wide and feathered, those of the female being mere jointed threads, without any feathering whatever. Many insects seem to be altogether without antennæ, but, like the undeveloped legs already mentioned, they can be found in their places, though so small as to escape a hasty observation.

HAVING now briefly examined the general characteristics of insects, we will take them in detail.

Among the insects, the COLEOPTERA, or BEETLES, are acknowledged to hold the first rank, their development being more perfect than is found in any of the other orders. The name of Coleoptera is composed of two Greek words, signifying sheath-wings, and is given to this order of insects in consequence of their leading peculiarity, which is, that the upper pair of wings is modified into horny or leathery cases, called elytra, useless in flight, but employed in protecting the membranous under pair of wings, which alone are used in flight. In many Beetles the lower pair of wings is not developed, and in a few both pairs are practically wanting, though an entomologist's eye can always detect them in a rudimentary form. The wings and elytra, together with other portions of the Beetle, will presently be figured and described.

The changes, or metamorphoses, of the Beetles, though singularly interesting, are not easily seen, as Beetle larvæ require food which is, as a rule, not easily procured, and in many cases is so noisome that few persons would like to meddle with it. A great number are carnivorous, feeding upon various living creatures, so that to supply them properly with food is next to impossible; while, as the majority of them pass two years or more in the larval state, the process of rearing them is tedious as well as difficult.

All who have bred silkworms, or have been in the habit of watching insects when at liberty, are familiar with the appearance of the three principal forms assumed by the moths and butterflies in their different stages of growth—the caterpillar, with its soft, cylindrical, ringed body, the pupa or chrysalis, covered with a hard, shelly case, and the perfect insect, with its beautiful wings. Now, although the Beetles pass through similar changes, they do not assume similar forms. The larva of the Beetle is, like that of the moth or butterfly caterpillar, soft and ringed, and sometimes so resembles the caterpillar that the two may easily be confounded by anyone unskilled in entomology. Figures of the larvæ will be given in connection with the different species.

Thus far the Beetle and moth bear some resemblance to each other, but when they come to assume the pupal form, they are so dissimilar that no one could possibly mistake one for the other. In the moth pupa all the limbs are hidden under a hard covering, but in the Beetle pupa all the members of the insect are visible, though they are covered with a skin which binds them down, and prevents them from being used until the insect attains its perfect form, and bursts through the investing skin.

If the reader will carefully examine the various stages of insect life, he will see that, whether the creature be in the larval, pupal, or perfect state, the number of rings of which it is composed are the same. Take, for example, a Beetle larva, and it will be found to consist of thirteen rings, or segments. The first of these forms the head the next three carry the legs, and the remaining rings constitute the body. Should the same larva be successfully reared, and examined after it has reached the perfect state, it will be found to have precisely the same number of rings, though some are fused together, and others are greatly modified.

IN the accompanying illustration we have an example of the Beetle, taken to pieces so as to show the various parts. The Stag Beetle has been chosen for the purpose.

We begin with the head and its appendages. Fig. 1 shows the mandibles, or upper jaws, which in the male Stag Beetle are of very great size. Fig. 4 shows the maxillæ, or lower jaws, with the maxillary palpi, or jaw-feelers, Fig. 4*a*. Fig. 3 shows the labium, or lower lip, with the labial palpi, or lip-feelers, 3*a*. Next come the antennæ, or feelers, Fig. 2.

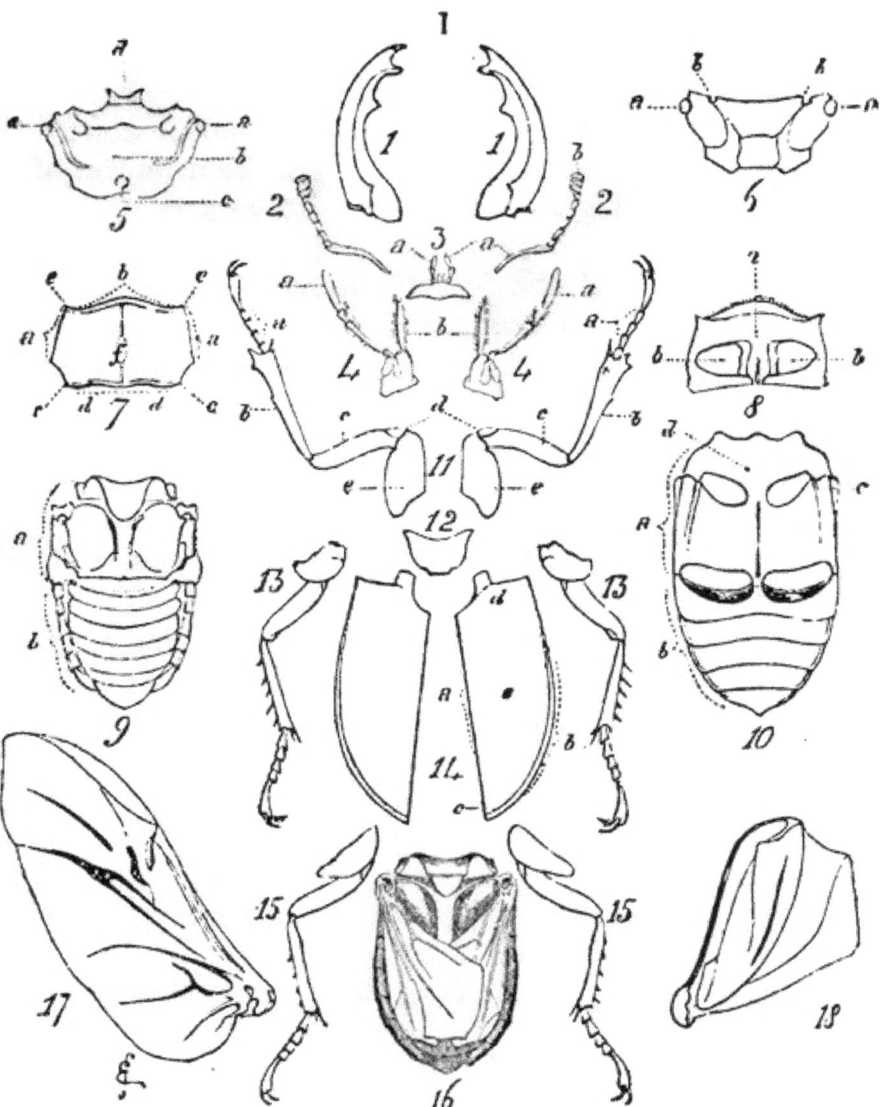

LUCANUS CERVUS. DISSECTION.

Parts of the Head.—1. Mandibles, or jaws. 2. Antennæ. 2*a*. Scape. 2*b*. Club. 3. Labium, or lower lip. 3*a*. Labial palpi, or lip-feelers. 4. Maxillæ, or lower jaws. 4*a*. Maxillary palpi, or jaw-feelers. 5. Head, upper surface. 5*a*. Eyes. 5*b*. Vertex, or crown. 5*c*. Occiput, or back of head. 5*d*. Clypeus, or shield. 6. Head, under surface. 6*a*. Eyes. 6*b*. Insertion of antennæ.

Parts of Thorax and Abdomen.—7. Pronotum, or upper surface of thorax. 7*a*. Lateral margin. 7*b*. Anterior margin. 7*c*. Posterior angles. 7*d*. Posterior margin. 7*e*. Anterior angles. 8. Prosternum, or under surface of thorax.

These organs are divided into two parts – namely, the scape, or long joint nearest the head, Fig. 2*a*, and the club, 2*b*. This latter portion is subject to most extraordinary variations of form, as will be seen in the following pages.

The head itself is shown at Fig. 5, the upper surface being here given. 5*aa* are the eyes, 5*b* the vertex, or crown of the head. 5*c* is the occiput, or back of the head; and 5*d* the clypeus, or shield, which covers the actual mouth. Fig. 6 shows the under side of the head. 6*aa* are the eyes, 6*b* the insertion of the antennæ.

As reference has been made to the eyes, it must be mentioned that these organs, although apparently only two in number, are in reality compound eyes, being made of a vast number of facets gathered into two groups, one on each side of the head. There is an apparent exception to this definition in the well-known Whirligig Beetles (*Gyrini*), which appear to have four eye-groups, two above and two below. This exception is, however, more apparent than real; the eye-groups being in fact only two, but each pair being crossed by a band of the horny material of which the outer skeleton is composed. The compound eyes can be seen to great perfection in some

8*a*. Sternum. 8*b*. Insertion of coxæ. 9. Meso-thorax and upper surface of abdomen. 9*a*. Mesothorax alone. 9*b*. Abdomen, upper surface alone. 10. Metasternum and abdomen. 10*a*. Metasternum alone. 10*b*. Abdomen, under surface alone. 10*c*. Parapleura, or side pieces. 10*d*. Episterna, or breast pieces. 12. Scutellum.

Legs.—11. Anterior, or first pair of legs. 11*a*. Tarsi, or feet. 11*b*. Tibia, or shank. 11*c*. Femur, or thigh. 11*d*. Trochanter. 11*e*. Coxa. 13. Intermediate pair of legs. 15. Posterior pair of legs.

The Wings.—14. Elytra. 14*a*. Suture. 14*b*. Lateral margin. 14*c*. Apex. 14*d*. Base. 14*e*. Disc. 16. Wings folded on abdomen. 17. Left wing expanded. 18. Right wing folded.

of the butterflies; but for this purpose, a careful manipulation of the microscope is needed.

In addition to these compound eyes, many insects possess several small simple eyes, called ocelli. They are very small, and in the Beetles are two in number, and placed on the back of the head.

Having examined the head, we now take the next division of the insect, namely the thorax. This important part bears all the instruments of locomotion, whether they be legs or wings; and is most wonderfully constructed for the purpose, being supplied in the interior with hard projections that are needed for supporting the powerful muscular apparatus needful for flight, and the less powerful, but still more important system by which the legs are moved.

The thorax is internally divided into three parts, which we call prothorax, or front thorax; mesothorax, or middle thorax; and metathorax, or hinder thorax. Beginning with the prothorax, we find it again divided into two portions, the upper and the lower—the former going by the name of pronotum, i.e. in front of the back, and the other called prosternum, i.e. in front of the breast. Fig. 7 shows the pronotum of the Stag Beetle: 7 *aa* are the lateral margins, 7*b* is the anterior margin, 7*cc* are the posterior angles, 7*d* the posterior margin, and 7*ee* the anterior angles.

Next we come to the prosternum, which is shown at Fig. 8; 8*a* being the sternum, and 8*bb* the insertion of the coxa, a joint which will be presently described.

The mesothorax with the abdomen is shown at Figs. 9 and 10, the former exhibiting the upper, and the latter the under surface. In the last figure, *a* is the metasternum, *b* the abdomen, *c* the parapleura, or side-pieces (sometimes called paraptera, because they are situated by the wings), and *d* the episterna, or breast-pieces.

Each of these portions is set apart for a definite use, and is employed for the attachment of some portion of the locomotive apparatus.

The prosternum is used to carry the front pair of legs, as can be seen by looking at the under side of any large Beetle, or indeed of any small one, by the aid of a magnifier.

The mesothorax bears the elytra, or wing cases, and the intermediate pair of legs, the former being attached to the upper part, or mesonotum, and the latter to the lower part, or mesosternum. The reader must remember that the word 'sternum' always signifies the breast, or under side of the thorax, and 'notum' the back, or upper side. Lastly, the meso-thorax bears the lower, or membranous, wings and the last pair of legs. As before, the wings are attached to the upper part, or metanotum, and the legs to the lower part, or metasternum.

We come now to the legs, the three pairs of which are represented successively at Figs. 11, 13, 15; the first pair being called the anterior legs, the second the intermediate legs, and the third the posterior legs.

These legs are divided into several portions, which are marked at Fig. 11. Beginning at the extremity of the limb, *a* is the tarsus, or foot, which consists of several joints. There are mostly five joints in the tarsus of Beetles, but in many families one or two of the joints are so small as scarcely to be visible, and only to be detected by a practised eye with the aid of a lens. Next comes the tibia, or shank, which is shown at *b*. Then follows the femur, or thigh, as seen at *c*. This is attached to a small joint called the trochanter, which is drawn at *d*; and last comes the coxa, or the joint which connects the limb with the thorax. The action of the coxa is very curious, it being a kind of ball-and-socket joint, but with a limited range of movement, so that the legs cannot be spread too far. The same divisions of the joints are found in the three sets of legs.

We next come to the wings. The upper pair, or elytra, are shown at Fig. 14. For convenience of description they are marked into several divisions. Fig. 14*a* is the suture, or line of junction between the two elytra. The apex is shown at *c*, and the base at *d*. The middle, or disc, is marked *e*, and the lateral margin is at *b*. Fig. 17 shows one of the wings expanded, as if for flight; Fig. 18 is the opposite wing, represented as folded, and at Fig. 16 are shown both wings as they appear when the elytra are removed.

At first sight some of these terms may appear to be harsh, repulsive, and difficult to master. In reality they are not so, and a knowledge of them is absolutely necessary to anyone who wishes to understand the description of an insect, and himself to describe insects intelligibly. They form a kind of shorthand by which knowledge can be rapidly communicated, and the trouble taken in learning them is amply repaid by the advantage gained by the student, even were the trouble multiplied tenfold.

But, in reality, there is scarcely any trouble needed. If the intending entomologist should content himself with merely learning a string of names by rote, he must expect to find his lesson a hard and repulsive one, and that it will be forgotten almost as soon as learned. Practical knowledge is ever the best, and the reader who intends to become an entomologist should take some Beetle—the largest he can find—and compare it, piece by piece, with the figures and description.

The most effective plan of all, however, is to take the Beetle entirely to pieces, and to lay out the portions on a card in their proper order, numbering each piece, and writing an index to the numbers. The various portions can be fixed to the card by gum tragacanth, which has the advantage of great holding power and dries without leaving the glittering surface which is found in most cements. I recommend taking two Beetles, so as to show the upper and under sides of each portion. This will be found peculiarly interesting in the thorax.

Before any attempt at dissecting the Beetle it should be steeped in soft water for a time, until the soft parts are thoroughly dissolved. The water should then be poured away, and fresh water substituted until the whole of the muscles and viscera have been washed away. Care must be taken lest any of the smaller joints be lost during this process. When the whole skeleton is fairly laid out, it can be mounted in a glass frame, and, besides serving as an infallible guide to the external anatomy of the Beetle, it is really a pretty and ornamental object. Many years ago, when I first began the study of entomology, I thus prepared several Beetles, and the knowledge thus gained has never been lost. Had I studied books alone, I should not have been able to gain the information half so easily, or to have retained it half as long.

CHAPTER II.

THE GEODEPHAGA.

THE word with which this chapter is headed is not a very alluring one, and yet to an entomologist it would say that the chapter contains the history of the best developed and some of the most interesting of the British Beetles. The term Geodephaga is formed from two Greek words, signifying earth-devourers, and is given to the large group of predacious Beetles which live on the ground. There is another large group, called the Hydradephaga, or water-devourers, i.e. those predacious Beetles which inhabit the water. We will take these two important groups in succession, selecting such examples as may best illustrate them.

It must be, in the first place, observed that any Beetle may be recognised as belonging to either of these groups by the structure of the mouth. In reference to the illustrations already described, the reader will see that each of the maxillæ (Fig. 4) is furnished with a four-jointed maxillary palpus. All these carnivorous Beetles possess the same organs, but, in addition, they have an inner lobe, which is also furnished with its palpus. Both these groups are associated in one large group called Adephaga, the word being a Greek one, and signifying greedy or gluttonous.

BEGINNING with the Geodephaga, we take the first family of the group, the Cicindelidæ, or Tiger Beetles. In all these Beetles there is a little movable hook at the end of the maxillæ, and the ligula is very short, and not appearing beyond the mentum. There is only one genus of these Beetles inhabiting England, and this is the typical genus Cicindela. The members of this genus vary but little in size, being about half an inch in length, having slender legs and antennæ, powerful and curved jaws, and very rounded and prominent eyes. They are all prettily coloured, and some are absolutely magnificent when viewed with the aid of a magnifying glass.

Never was a popular name more appropriate than the title of Tiger Beetles, which has been given to this genus. If we can imagine tigers who, in addition to active limbs, their teeth, and their talons, are furnished with large and powerful wings, we can form some idea of the part which these creatures play in the world of insects.

We will take as our typical example the common Green Tiger Beetle or Sparkler (*Cicindela campestris*).

The colour of this beautiful beetle is gold-green above, and shining copper-green below; and there are several yellowish spots on the elytra, varying much in shape, number, and hue. Sometimes there are only three, but in many specimens there are six.

In former times, the variety in the number of spots was thought to indicate that the Beetles belonged to different species, but it is now decisively ascertained that they are only varieties of one single species.

If the wing-cases be opened, and the broad membranous wings spread, the upper surface of the abdomen is seen to be deep shining blue, very much like the colour of the ordinary 'blue-bottle' fly. As the Beetle darts through the air in the sunshine, the light glitters on the burnished blue surface, a circumstance which has earned for the Beetle the popular name of Sparkler.

One peculiarity of this insect is the strong but pleasing scent which it emits. I well remember the first time that I saw and captured this Beetle. It was on a sandy bank in Bagley Wood, near Oxford, and I could not for some time guess the origin of the pleasant, sweet-briar sort of scent which clung to my hands. For some time I thought that I must have grasped some fragrant herb, and it was not until I had taken the Beetles out of the box (where, by the way, nearly one half had been killed and partly eaten by the other half) that I discovered the real source of the perfume.

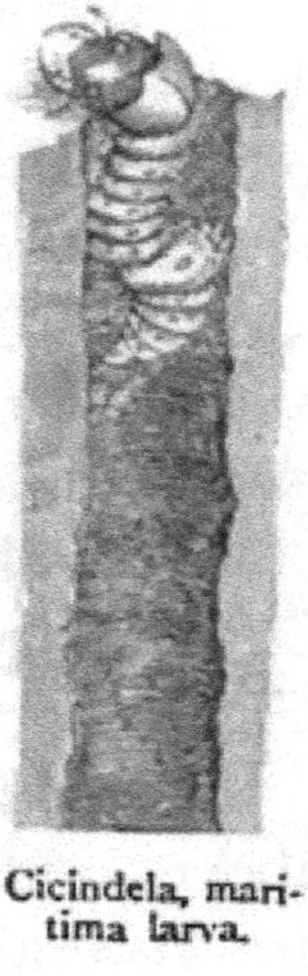

Cicindela, maritima larva.

With all the Tiger Beetles, the larva is an odd-looking grub, with a pair of enormous, sickle-shaped jaws, and a sort of hump on the middle of the back. These larvæ make perpendicular burrows in the ground, and lie in wait for passing insects. Sandbanks, if partly overgrown with grass, are favourite localities of the common Tiger Beetle.

We now come to the next family of British Beetles, the Lebiadæ. All the Beetles of this family may be known by the appearance of the elytra, which do not extend to the end of the abdomen, and are abruptly squared, looking almost as if they had been cut off. If their first pair of legs be examined, the tibiæ will be seen notched on the inner side. They are all rather small Beetles, and some are very prettily coloured. They are tolerably active when they choose to take the trouble of moving; but they are much given to hiding themselves in all kinds of crevices, so that some trouble is requisite in order to procure them.

The young entomologist must always bear in mind that the most unpromising localities will often prove to be singularly rich in insects, and that Beetles especially may be found in any spot where there is a crack or a hole. Large stones are nearly sure to shelter a Beetle or two beneath them; moss is generally full of them; and a heap of decaying grass or leaves is a hotbed which seems as well fitted to produce Beetles as to force plants. The loose bark of trees always has Beetles under it; and small Beetles creep into the burrows which larger Beetles have made in the decaying wood of the tree. When a quantity of long moss is to be seen, it is a good plan to fill a bag with it—a paper bag will do in lieu of anything better—and to bring it home, when it can be carefully examined by shaking it bit by bit over a large sheet of white paper. Grass-tufts can be treated in a similar manner, and mostly with great success. Lumps of dry earth can also be brought home, where they can be broken up and leisurely searched.

THERE is a common species of this beautiful genus to be procured from the broom. This is *Lebia chlorocephala*. It is a striking insect in appearance, the head and elytra being of a brilliant bluish-green colour, while the thorax, which is strongly punctured, is of a rust-red hue. It is very variable in size, some specimens being twice as large as others, but averages about a quarter of an inch in total length.

THE genus Dromius comes next on our list. There are eleven British species of Dromius at present known, all of them small and rather pretty insects. Although they are almost invariably found under the bark of trees, they must not be confounded with those little Beetles which devour the bark or bore into the wood. On the contrary, such insects constitute the food of the Dromii, so that the latter ought to be encouraged and protected as far as possible. Their bodies are long and flattish, so as to enable them to run about under the bark in search of prey.

In the accompanying illustration is shown a Beetle which may be accepted as the type of the genus. Its name is *Dromius quadrimaculatus*, the latter name being given to it on account of the four white marks upon the elytra. The head of this species is black, and the thorax is rust-red. The elytra are brown, and each of them has two white spots, as shown in the illustration. The length of the Beetle is designated by the line drawn by its side. It can be taken in various localities. Throughout the year it can mostly be found by removing the dead bark of trees, but in winter it can sometimes be found under stones and in heaps of decaying leaves.

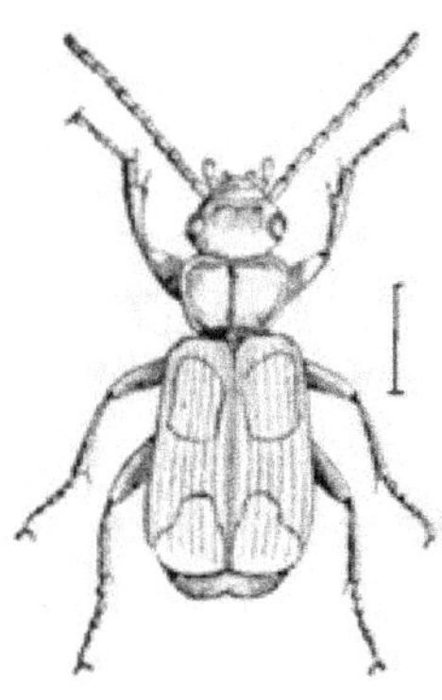
Dromius quadrimaculatus.

The word Dromius is of Greek origin, signifying a runner, and is given to the members of this genus in consequence of their activity.

THERE is a prettily-made, though soberly coloured little Beetle, called *Brachinus crepitans*, the latter title being given to it in consequence of a most singular property which it possesses, and which is almost unique in the animal kingdom. It secretes a remarkable volatile fluid, which it has the power of retaining or expelling at will. When alarmed, it throws out a small quantity of this fluid, which immediately volatilises with a slight explosion when it comes in contact with the atmosphere, and looks very much like the fire of miniature artillery. In consequence of this phenomenon, the insect which produces it is popularly called the Bombardier Beetle.

This curious property is used in defence. The Beetle, being a small and comparatively feeble one, is liable to be attacked by the larger Geodephaga, especially by those belonging to the genus Carabus. The lesser insect could have no chance of escape but for its curious weapons of defence. When the Carabus chases the Brachinus, the latter waits until the former has nearly reached its prey, and then fires a gun, so to speak, in its face. The effect on the Carabus is ludicrous. The insect seems quite scared at such a repulse, stops, backs away from the tiny blue cloud, and allows its intended prey to reach a place of safety.

Carabus monilis. Brachinus crepitans.

The Brachini may be identified by the very convex body, and their palpi without the axe-shaped terminal joint. The Bombardier Beetle is reddish-yellow, with dull deep-blue elytra. It is, however, a variable species, as are all, or nearly all, of those in which green or blue is the prevailing colour, and, though most specimens are blue, or blue-black, some are deep blackish-green. It also varies greatly in size, some specimens being not a quarter of an inch in length, and others more than the third of an inch long.

These Beetles love wet situations, especially when the water is brackish, and hide under stones and in crevices, so that they are seldom seen except by insect hunters. The banks of tidal rivers are good hunting grounds for the searchers after Brachini, such as the Thames, from Erith, or even Woolwich, to its mouth. They are found in greatest numbers below

Gravesend, and ten or twelve may sometimes be seen under a single stone, firing off their artillery when deprived of their shelter.

The volatile fluid which produces such curious effects is secreted in a little sac just within the end of the abdomen. It is not only capable of repelling the larger Beetles by its explosion and cloud of blue vapour, but is potent enough to discolour the human skin when discharged against it, as many have found who have captured Bombardier Beetles by hand. Should it get within the eyelids, the pain and irritation produced resemble those which would be caused by a corresponding amount of the strongest vinegar.

The whole of the contents are not ejected at one discharge, but there is sufficient to produce a series of explosions, each perceptibly fainter than its predecessor. Even after the death of the Beetle, the explosions may be produced by pressing the abdomen between the finger and thumb. Even in our small British species the phenomenon is very surprising, but there are much larger species in hotter countries, which produce much louder explosions, accompanied with quite a cloud of vapour. Three British species of Brachinus are known to entomologists, the remaining two, however, being extremely rare.

We come now to another family of Geodephaga, of which we can but take one example. The Scaritidæ, like the preceding family, are seldom seen in the open air, but, instead of creeping into clefts already existing, they make tunnels for themselves. Tunnelling Beetles are almost invariably cylindric in shape, and this is the case with the Scaritidæ. The thorax of these beetles is rather peculiarly shaped, being separated from the abdomen by a sort of neck, or 'pedunculated,' if we use the scientific term. The tibiæ of the first pair of legs, which are the tools chiefly used in burrowing, are very hard, very strong, and boldly toothed, and the antennæ are short, and have scarcely any distinction between the joints.

One of the best examples of the Scaritidæ is shown in the accompanying wood-cut, and is an insect called *Clivina fossor.* The length of this Beetle is rather more than a quarter of an inch. Its colour is exceedingly variable, ranging from pitchy-black to chestnut, brown, or even brick-red. This variation in colour depends chiefly upon the exposure to the air, the oldest specimens being the darkest. This gradual darkening by exposure to light is very frequent among insects; and a too familiar example may be found in the common cockroach, which is often seen almost white, darkening gradually until it assumes its well-known red-brown hue. There are only two British examples of this genus.

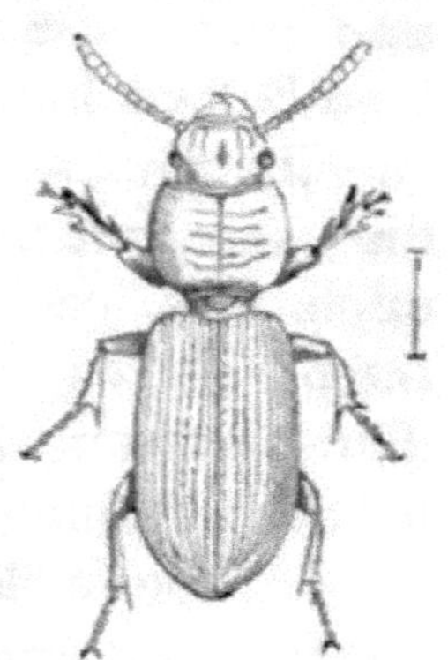

Clivina fossor.

Usually, the Clivina is found under stones and in heaps of decaying vegetable matter, but I have taken great numbers out of a large rotten log, which was seen to be full of their burrows as soon as the bark was removed. I took out of the same log both the larvæ and pupæ of the same Beetle, having nothing to do but to break up the soft and rotten wood with a powerful digger, previously laying a newspaper below the log. The insects fell out in numbers without being injured, and could be taken in any number.

During the winter-time the Clivina is usually to be obtained by digging at the roots of trees, and carefully breaking up the sods.

When the insects are captured, the next business is to know how to deal with them. There is not the least difficulty with soft-bodied moths or butterflies, as they can be easily killed when caught; but the hard-bodied Beetles are not so easily deprived of life, and a great number of them will fight if placed in the same vessel, and eat, or at least mutilate, each other. There is nothing better for such insects, or rather for their capturer, than the 'laurel-bottle.' This is very easily made. Take a wide-mouthed bottle, and fit a cork very firmly into it. Bore a hole through the cork, and insert in the hole a swan-quill or short metal tube about two inches long, so that it may project at least half an inch through the cork into the bottle. The upper end should be stopped with a cork, and it will be better to cut the cork so long that it can be removed by holding it between the teeth. I always cut the upper part of the tube slopingly, so that a very small Beetle can be scooped up with it.

For many Beetles nothing more is requisite except to put a piece of crumpled rag at the bottom so as to give them a foothold; but for the fiercer and voracious Adephaga an additional precaution is required, and they must be killed as soon as they are put into the bottle, or an undisfigured specimen will never be obtained.

If a bottle can be obtained without any neck, the following is the neatest way of making a laurel-bottle:—Cut a flat cork that fits rather tightly in the bottle, but not too tightly to be pushed up and down. Take out a portion of its middle, and insert a piece of fine wire gauze. It will be better to pass a string through each side of the 'plug,' as we will call this cork, and knot them underneath, so that when the plug is to be withdrawn it can be pulled up by the strings.

Next, procure a handful of the young buds and leaves of laurel, put them into the bottle, and crush them into a paste with the handle of a knife or some such instrument. Now insert the plug, press it down upon the crushed leaves, put in the cork, and the 'laurel-bottle' is complete. The bottle must be kept firmly corked, or the vapour will escape.

The use of the bottle will be shown as soon as it is employed. Let a Beetle, however large, fierce, or voracious, be put into it, and its fate is at once sealed. It begins to kick and struggle, as if it knew its danger; but in a very few seconds the struggling is evidently over, and the insect turns on its back, with its legs quivering in the poisoned air. The fact is, the laurel contains a large amount of prussic acid, and the interior of the bottle is charged with its vapour. Now, as has already been mentioned, the breathing apparatus of an insect pervades the whole of the body, even to the end of the limbs; and when the poisonous vapour is inhaled, it penetrates simultaneously the entire system, and causes almost instantaneous death. It will be as well for the beginner to use the laurel-bottle for all Beetles, as it does not damage them, and he need not trouble himself to distinguish the voracious from the harmless species.

There is only one drawback to the laurel-bottle, and this is more apparent than real. It stiffens the limbs at first, so that the insects cannot be 'set' properly, and the legs will rather break than bend. However, this difficulty is overcome in a very simple way—namely, by leaving them in the bottle for a few days, when the rigid limbs will become relaxed and as flexible as they were during life. Some care, however, is required, as, if they are left too long in the bottle, the process of softening extends too far, and the limbs are apt to fall off altogether.

Very few instruments are required for the capture of Beetles besides the digger which has already been described. Two nets are almost indispensable, one made of stout net, and the other of brown holland or even canvas. The former should be about seven inches in diameter, and is used for fishing Water Beetles and their lárvæ out of ponds, ditches, and streams.

The latter, which is called from its use the sweep-net, requires to be made with some care, as it meets with very rough usage, and, unless properly made, will soon be worn out. The depth should be at least twice its diameter.

The framework of the net is simply a ring or hoop about nine inches in diameter, made of iron about one-third of an inch in diameter. If it be of lighter material, it will not endure the rough work for which it is made. The ring is covered loosely with stout leather, and to the leather is sewn the net itself, which is nothing more than a bag of holland or canvas. The best sweep-nets have a number of small rings forged on the inside of the hoop. A stout wire is run through the rings, and the net is made fast to the wire. The socket and ring must be *welded* together and not soldered. The angles at the bottom of the net should be carefully rounded off. The handle should be made of ash, stout, and about five feet in length.

Now for the mode of using it. When a promising sweeping-place is found, such as a hedge-row, some long grass, fern or heather, a quantity of nettles, a turnip field, or any such locality, the sweep-net is swept at random backwards and forwards among the herbage, the stroke always having an upward direction. This, with a little practice, can be done so rapidly as scarcely to impede the walk. It is better, however, only to sweep one kind of plant at a time, so as to be sure of the particular herb or vegetable frequented by each species.

After a time, comes the examination of the net. Lay it flat on the ground, doubling it over, so that the pressure of the hoop prevents the inmates from escaping. Now, draw it through the hoop very slowly, taking care that none of the more active Beetles make their escape, especially those which hop and fly; seize the insects as they make their appearance, and drop them into the laurel-bottle, always replacing the cork.

Most Beetles—and, in fact, nearly all that are taken in the sweep-net—will pass through the tube; but the large Ground Beetles, some of the Water Beetles and chafers, and one or two others, require to have the large cork removed.

On returning home, the entomologist should take out the cork from the laurel-bottle, and empty the insects into some boiling water, as some of the species have the strongest objection to die, and, after they have been apparently killed, have a habit of reviving in a manner that is rather startling to the young entomologist. It is no uncommon thing for an insect-hunter to capture a number of Beetles, set them, put them away in the 'setting-box,' and then, after a week or so, to find three or four of them kicking about and doing their best to escape. Some of the Geodephaga have been known to drag the pin from the board, and, though still impaled, to devour their fellow-sufferers.

Ordinary Beetles can be taken by hand, but for the very small species the forceps is used. This instrument should be of good length. The regular dealers offer for sale a brass forceps about two inches in length. This is nearly useless. Get the forceps of steel, at least four inches long, and made with a curve. The curved forceps is just as useful as the straight instrument for picking up a Beetle from the ground, while it can be inserted into crevices which the straight forceps could not enter.

'Setting' Beetles is not a difficult matter. For moderately-sized Beetles the following plan answers perfectly. Take a fine pin, such as are sold for the purpose, and pass it through the disc of the right elytron, and so fix the Beetle on a board. Now draw out its legs, and place them in their natural position, fixing them, if required, with pins and little bits of card-board. Treat the antennæ and palpi in the same way, and nothing more is needed.

The smaller species should be fixed on strips of white card-board by means of gum tragacanth, of such a consistency that it will scarcely flow over if the bottle containing it be held upside down. Only one Beetle should be set on each piece of card. The legs, antennæ, part of the mouth, &c., must be drawn into position by means of the forceps and the setting-needle. At first this operation will be found a matter of considerable difficulty, but practice and patience will soon enable the collector to set his capture with ease and rapidity. When the insects are mounted, an entomological pin should be passed through each strip at the end opposite to that on which the Beetle is placed, and the specimens transferred to the box or cabinet in which they are permanently to remain.

For examining the details of a Beetle, especially if it be a small one, a pocket lens is required. These instruments are made with either two or three glasses, and are small enough to be suspended to the watch-chain by a ring. In order to examine a Beetle with ease, the pin should be stuck into a cork cemented on a flat piece of lead, so that it cannot be knocked over. My own instrument is made of a champagne cork, cut into a cylindrical form and rounded on the top.

I prefer the champagne cork for two reasons—the first being that it is of an uniform and close texture, without the holes and hard spots which are found in ordinary corks, and the second being that it has not been pierced with the corkscrew.

CHAPTER III.

GEODEPHAGA—*continued.*

WE now come to the important family of the Carabidæ, to which belong some of the largest and most powerful of the Geodephaga. The Carabidæ are in many points exactly like the Cicindelidæ, and if isolated parts of the mouth were taken to an entomologist, he would have some difficulty in knowing to which family they belonged. But, whereas the Cicindelidæ have a notch on the inner side of the front tibiæ, these limbs are without the notch in the Carabidæ.

The typical genus is well represented in this country, and its members are familiar to us by the title of Ground Beetles or Garden Beetles. They are the largest of the family, some of them being an inch in length, and strongly though elegantly shaped. They are very active, as far as their legs go, but they have no wings, these members being only found in a rudimentary state under the hard and shining elytra, which in most of the species are soldered together and cannot be opened. In one species, however, *Carabus granulatus*, the elytra are capable of motion, and the wings are more developed than is generally the case. They are, however, much too small to be used for flight. The mandibles have a small tooth in the middle, and the labial palpi have the last joint 'securiform,' or axe-shaped. The thorax is somewhat heart-shaped, and has the posterior angles boldly marked. In the male the tarsi of the first pair of legs are broader than in the female.

Thirteen British species of Carabus are known, one of which is represented on page 23, in the act of chasing the Brachinus. It is the *Carabus monilis*, a common and very handsome insect. It is exceedingly variable in colour, and slightly so in marking, but may be described as follows:—The head is black, with bronze or green reflections, wrinkled in front, and there is a deep hollow on each side just between the antennæ. The thorax is wrinkled at the hinder angles and deeply notched behind, and its colour is deep copper. The elytra are metallic green or violet, and sometimes entirely green. Each elytron is marked with three rows of raised striæ, broken regularly at intervals, or 'interrupted,' to use the scientific term, and between them are three ridges which are not broken. The line along the suture is black, and the under side of the insect is dull black.

This Beetle is common in gardens, and, like the rest of its kin, ought to be encouraged and protected, as it feeds almost entirely on the smaller insects, and never meddles with the vegetables.

Another very good example is *Carabus violaceus*, one of the commonest and largest of the genus. It sometimes exceeds an inch in length. Its colour is rather remarkable. At first sight it appears to be a black insect, but a more careful examination shows that the supposed black is in reality the deepest violet, which on the margins of the elytra and thorax becomes of a rich burnished golden-violet, like as of polished metal. The whole of the upper surface is granulated, the elytra more deeply than the thorax, along the centre of which there is a slightly elevated line. Below, it is black, with a slight green or blue reflection.

These Beetles afford good practice in setting. They are large enough to bear handling, and yet small enough to require care. Their legs are long, and look well when set out, and the parts of the mouth are sufficiently large to show whether the operator has been careful about his work. These Beetles, by the way, are very tenacious of life, and, though they can be at least rendered insensible and harmless by the laurel-bottle, it will be as well to dip them into boiling water before passing the pin through them, so as to avoid the sight of an impaled Beetle trying to release itself, or walking about the cabinet drawer with a pin through its body.

I have always had a great liking for these Beetles from the time when I was accustomed to harness them into fairy chariots, to that in which I first learned from them the wonders of an insect's organisation and traced in them the early rudiments of those structures which find their fullest development in man.

NEXT on our list comes the genus Notiophilus, which, being translated, signifies wet-lover and is a very appropriate title. All these Beetles are very small, none of them exceeding a quarter of an inch in length. They may be found on the banks of ponds and streams, running over the wet soil with great activity in search of prey. In consequence of their fondness for water they were for a long time classed among the Hydradephaga, or the voracious Beetles of the water, but have now been placed in their right position among the Carabidæ.

Accompanying this description is a magnified figure of a common species of this genus, called *Notiophilus biguttatus.* It is a very small Beetle, sometimes only one-sixth of an inch in length, and seldom, if ever, exceeding a quarter of an inch. It is a pretty little creature, with a highly polished surface, as if made of bronze in which the copper predominated. Five species of this genus are known to inhabit England, and they may be found hidden under stones or at the roots of trees in damp places. Willow trees on the borders of streams are good localities for these pretty little Beetles. They do not, however, confine themselves to wet places, although they prefer such places as a residence. They may be seen running about in the hottest weather over places a mile at least from water.

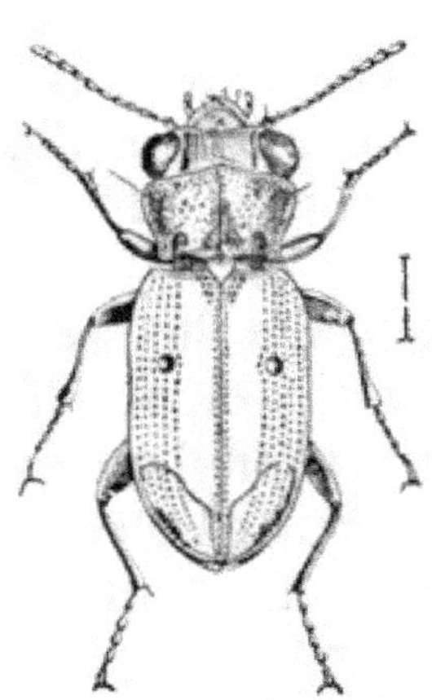

Notiophilus biguttatus.

NEXT upon our list comes the family of the Chlæniidæ. In these Beetles the front tarsi of the males have either two or three joints much widened and squared, and very sponge-like below.

A well-known beetle belonging to this family is shown. Its name is *Pristonychus terricola*. As may be seen by reference to the illustration, it is a very prettily-shaped insect, the curves of the outline harmonising in a way that would have delighted the soul of Hogarth, had he taken the trouble to look at Nature's original of his celebrated 'line of beauty.'

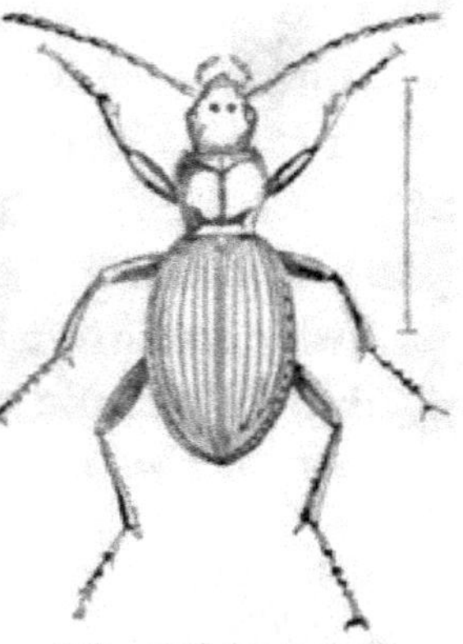

Pristonychus terricola.

Except in shape, it is not a very pleasing insect to the eye, the colour being black, with a violet gloss when examined in a proper light. The head is pitchy-black, and the thorax is sometimes of the same colour as the head, and sometimes blue-black, with a faint furrow in the centre, and a deep oblong pit on either side of the base. The elytra are of the same colour, and rather flattened, and covered with faint but regular striæ, which are slightly punctured. Along the edges there is a series of roundish pits.

This Beetle is plentiful in most parts of England, and may be found in cupboards, cellars, dark out-houses, and similar localities. There is only one British species.

WE now come to the pretty little Beetles that are ranked under the generic title of *Anchomenus*. All these insects have an elongated thorax, the head egg-shaped and the antennæ with the third joint twice as long as the second.

They are very active, and very brilliantly coloured, and, like most bright-coloured Beetles, love the sunshine, in which they dart about with exceeding rapidity. The popular name of Sun Beetle is given to these and other insects, in consequence of this peculiarity. They are sociable little creatures, and, when one is seen, others are tolerably sure to be close at hand. Some of them frequent wet and marshy places, and may generally be found at the roots of willows that are planted at the water-side.

One example of this interesting genus is *Anchomenus dorsalis*, which is shown, rather magnified, in the accompanying woodcut. The real length of the insect can be known by reference to the line that is drawn on its right side. In this species the head and thorax are dark-green, and the flattened elytra are pale rust-red, diversified with a large spot of blue-black towards the apex, but not quite reaching the tip. They are striated, and the interstices between the striæ are flat and smooth, without any punctures. Beneath, it is shining black.

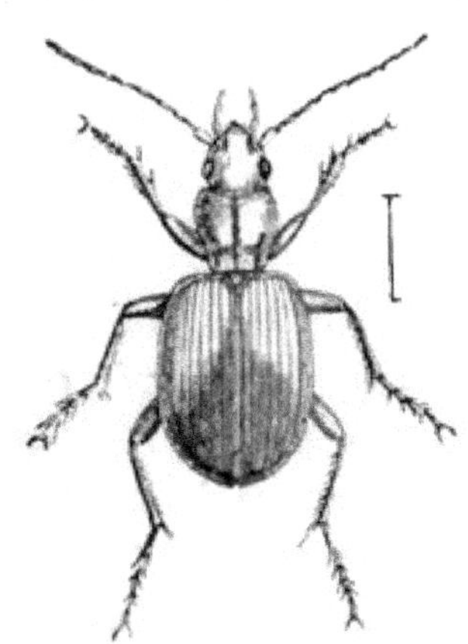

Anchomenus dorsalis.

THE family of the Feroniidæ, which comes next in order, comprises a number of Beetles, none of which are remarkable for beauty, though there are several whose habits are very curious and interesting. They may be known by the sinuated, or wavy, form of the elytra at the apex, and by the basal joints of the front tarsi of the males. These joints are not squared, like those of the Chlæniidæ, but are somewhat heart-shaped, and furnished with two rows of bristles beneath.

Our first example of this family is the Beetle which is known to entomologists by the name of *Pterostichus madidus*, a figure of which is given herewith. The colour of the insect is shining black, with a slight brassy gloss. The smooth head has two impressions in front; the thorax is convex and narrow behind, with a bold central furrow and a deep wrinkled pit at each angle. A lens is required to make out these details. The elytra are covered with regular striæ, a circumstance which has gained for these insects the generic title of *Pterostichus*, or 'streaked-wing.' There are a few small punctures on each elytron, and a row of bold punctures runs along each margin. The wings are not developed.

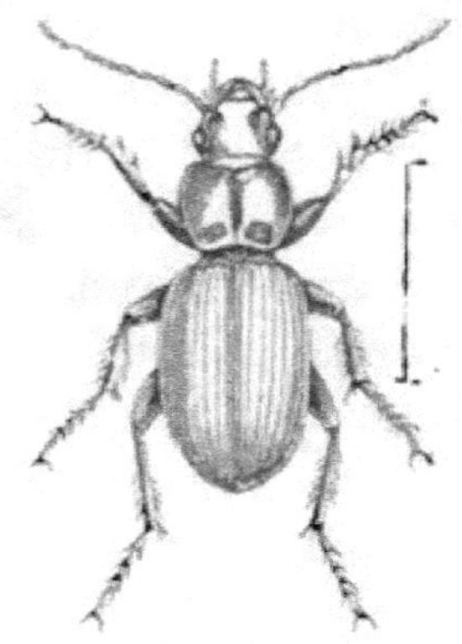

Pterostichus madidus.

This is a very common insect, and can be taken in any number, as it runs boldly about. Anglers often use it successfully as bait. Although devoid of wings, it is very quick on its legs, darting about with such velocity that the generic name of *Steropus*, or 'lightning,' was once applied to it and another allied Beetle.

Twenty-two British species of this genus are now acknowledged, but in it are merged several genera of the older entomologists.

WE now come to one of the largest and most interesting insects of this family, though assuredly it is not a beautiful one. This is *Broscus cephalotes*, an insect which has been called the giant of its family. This name it well deserves, as it rivals the Carabi themselves in size, sometimes reaching nearly an inch in length. Being a predacious Beetle, it is gifted with very powerful jaws, which are attached to a head of more than ordinary size. The specific name *cephalotes* signifies large-headed, and is appropriately given to this Beetle. The generic name *Broscus* signifies a devourer, and is equally appropriate.

This insect can at once be known by the very large head, and the shape of the thorax, which is much narrowed towards its base. A deep furrow runs along its centre, and its base has a deep pit on either side. The elytra are rounded at the shoulder, and striated; some small punctures being scantily visible on the striæ. The colour of the insect is black.

In this Beetle we see one of the fiercest and most voracious of the whole insect race. It lives on the sea-shore, generally hiding itself beneath decaying sea-weed or stones, and making burrows under such points of vantage. From this burrow it issues in search of prey, and successfully pursues all kinds of insects, its own kind included. So voracious is it, and so many insects does it kill, that if it reside for a day or two in one burrow, it can be detected by the rejected elytra, limbs, and other parts of insects which it has caught and eaten. It is the only British example of its genus.

THE large genus Amara now comes before us, and out of the twenty-four species which are included in it I have selected *Amara obsoleta* as our example. This insect is depicted in the accompanying woodcut. All the insects belonging to this genus are small, and most of them are brightly coloured. They all take rank as Sun Beetles or Sun Shiners; and, fortunately for them, there is a wide-spread superstition that it is unlucky to kill a Sun Beetle, and that its death will cause terrible storms.

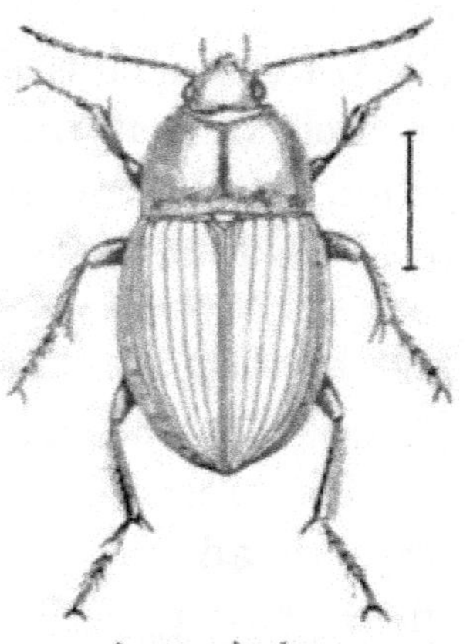

Amara obsoleta.

The members of this genus are rather wide in proportion to their length, and have the thorax wide behind, as wide, in fact, as the elytra. They have large wings, which they can use with great effect; and the males have three dilated joints on the front tarsi. These Beetles are very plentiful, and may be seen either flying through the air on their ample wings, running about in full blaze of the sunshine, or temporarily hiding under sticks and stones.

Although it is no very difficult matter to know an Amara when it is seen, I must warn the reader that to distinguish the different species is a task which requires the minutest attention to the smallest details, and had better be deferred until the eye has been trained to seize at once on those small but important characteristics, which at once strike the eye of a practised entomologist, and invariably elude the scrutiny of a novice. The eye can only see that which it has the power of seeing; and it is worthy of remark that twenty or thirty young observers will miss exactly the most important detail in an insect structure until it is pointed out by an experienced entomologist, when they will at once see it, and wonder how anything so obvious could have eluded them.

In any large genus of insects there is always a difficulty in deciding upon the different species; and, even among the moths, where size and colour are tolerably constant, mistakes are continually made. But, among Beetles, these important elements of size and colour go for almost nothing, and whenever green and blue are in question, colour absolutely does go for nothing. Now, in the members of the genus Amara, blue and green are the leading hues; and five individuals, which undoubtedly belong to the same species, may be respectively bluish-green, greenish-blue, brassy, coppery, or even black. Then the head and thorax may be of one colour, and the elytra of another; so that no dependence can be placed upon so uncertain a characteristic.

The present species is a very common one. Its colour varies from green to black, glossed with brass. The elytra are striated, and the striæ are bolder and deeper towards the apex than at the base. The head is nearly smooth, but has a few striæ drawn across it in front.

WE next come to the large family of the Harpalidæ, of which we shall take one example illustrating the principal genus. The males of the Harpalidæ have the four basal joints of the front tarsi dilated, and sometimes the corresponding joints of the middle pair of legs. These dilated joints are covered with stiff bristles. The mentum is deeply notched, and has a small lobe in the centre.

Unlike the preceding family, which are for the most part lovers of light and fond of darting about in the full radiance of the sunbeams, the Harpalidæ withdraw themselves from the light, and hide themselves during the daytime in any crevice that may present itself. Should, for example, the season be a dry one, the cracks in the earth are sure to be tenanted by Harpalidæ; and when the spade is employed, many of the Beetles are turned up together with the soil in which they have sought a refuge, and sought it in vain.

Our typical example of this interesting family shall be the very plentiful insect scientifically known as *Harpalus æneus*. In this genus three joints of the tarsi of both the front and middle pairs of legs are dilated in the males.

The pretty species which has been chosen as our example is polished on its upper surface like a mirror, the colour being exceedingly variable. Some specimens are brassy, others coppery, others green of various shades, and others again blue of various shades, deepening into violet so dark that it appears to be black. There are some faint striæ on the elytra, and in the flat interstices between the striæ are punctures, very few towards the suture, and plentiful towards the margin. Beneath it is pitchy-black. In this insect, as indeed in most of the members of this genus, the females are much duller than their mates, this effect being produced by a very fine granulation of the elytra. There are more than thirty known British species of this genus.

WE now come to the last family of the Geodephaga, namely, the Bembidiidæ. In all these Beetles the palpi are formed differently from those of the preceding families. If the parts of the mouth be carefully observed, the last joint but one both of the maxillary and labial palpi will be seen to be very large, while the last joint is very short and very small, so small indeed that at first sight it looks more like a spur than a separate joint. All these insects are lovers of salt and wet places, and are found on salt marshes near the mouths of tidal rivers, such, for example, as those which cover the district between Rochester and Sheerness, and upon the sea-shore itself.

Small as they mostly are, they are exceedingly voracious, and can kill creatures much larger than themselves. There is, for example, *Cillenium laterale*, a little copper-coloured Beetle, which never exceeds one-sixth of an inch in length and is generally much less, which, in spite of its small size, feeds on the common sand-hopper, seizing the active crustacean under the body and so destroying it. Like the Æpys, which has already been described, this insect passes much of its time submerged under salt water.

OUR first example of this interesting family of Beetles is taken from the typical genus, and is called *Bembidium biguttatum*.

Its colour is brassy or bronze-green, and its surface is polished and shining. The head has a shallow impression on each side. The thorax has a slight furrow along the centre, and a depression near each basal angle. The elytra are striated and punctured nearly as far as the apex, and between the second and third striæ there are two bold impressions, from which the insect derives its specific name of *biguttatum*, or 'two-channelled.' There is a reddish-brown spot at the apex. The under surface of the body is black, glossed with brassy or bronze reflections.

OUR last example of the *Geodephaga* is the insect called *Bembidium quadriguttatum*, which is shown in the illustration accompanying this description.

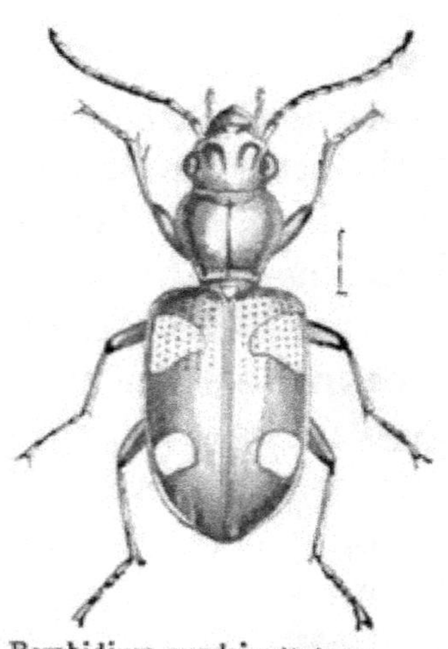

Bembidium quadriguttatum.

This very conspicuous little Beetle is tolerably common, and may be found in most damp places, whether the water be fresh or salt. The smooth head has a deep but short furrow on each side. The thorax is remarkably convex in front, and its colour is shining blue-green, or greenish-blue, as the case may be. The elytra are rather convex, and their colour is something like that of the thorax, but deepening into violet, which is sometimes so dark that it appears to be black. On the shoulder of each elytron is a patch of cream-white, and there is another near the middle, the insect deriving from these white marks the specific title of *quadriguttatum*, or 'four-spotted.' The spot on the shoulder is usually rather triangular, and that on the middle of the elytron nearly round.

CHAPTER IV.

HYDRADEPHAGA.

WE have now completed our notices of the Geodephaga, the analogues of the land Carnivora among the higher animals, and we now come to the HYDRADEPHAGA, or carnivorous Beetles of the water—the whales, porpoises, and seals of the insect world.

We know that all animals are specially adapted to the life which they have to lead, and therefore may naturally expect that Beetles which live in the water will be formed very differently from those which reside on the land, even though that land be constantly wet. Shape, for example, is likely to be altered. We know that the whales, dolphins, and seals, who have to pass either the whole or the greater part of their lives in the water, and to catch in it the living prey on which they feed, become assimilated in shape to the fishes; and it is likely that insects will obey the same laws as mammals. This is really the case, the shape of all the Hydradephaga being very fish-like, in order to enable them to pass more easily through the water. As there is much more friction in passing through the water than through the air, the Water Beetles, as the Hydradephaga are familiarly called, have the various portions of the body fitting closely to each other, so as to leave an uniform smooth and polished surface, something like the scaly surface of the fish, the slippery skin of the whales and dolphins, and the close-set fur of the seals.

The limbs are also modified to suit the special purpose for which they are designed. As these Beetles walk less than they swim, greater provision has been made for the latter mode of progression. Accordingly, the first and middle pairs of legs are comparatively small and feeble, the strength being thrown into the hinder pair, which are large in comparison with the others, and, in nearly all cases, flattened and furnished with a fringe of stiff bristles on the inner side, so that they serve as oars. They are jointed in a peculiar manner to the body, so that there is room within the thorax for a set of very powerful muscles which work them, and they are placed farther back than is usual among Beetles—a peculiarity of structure which is found also in the seals and the diving birds, especially the penguin tribe.

Although they cannot, as a rule, walk well, they can all fly well, and are furnished with very large and powerful wings, so that, if food should fail them in one piece of water, they can fly to another. They generally fly at night, and have an odd way, when they reach a pond or stream, of closing their wings while high in the air, and allowing themselves to fall like stones into the water. Sometimes, deceived by the glitter in the moonshine, they have been known fall upon the roofs of greenhouses.

Not only the Beetles, but their larvæ inhabit the water, and they are equally predacious in both stages of existence, the larva being armed with a pair of enormous sickle-shaped jaws. They are all long and narrow, and have six minute eyes, or ocelli, at each side of the head. We will now proceed to our examples of these insects.

THE Hydradephaga are divided into two families, the Dyticidæ and the Gyrinidæ. There is not the least difficulty in deciding the family to which any Water Beetle belongs, as a glance at the antennæ is sufficient for the purpose. The antennæ of the Dyticidæ are long and slender, and those of the Gyrinidæ are stout, short, and club-like. Moreover, the first pair of legs are short in the Dyticidæ and long in the Gyrinidæ. We begin with the first family, and take an example of the typical genus.

The GREAT WATER BEETLE (*Dyticus marginalis*) may be found plentifully in almost any pond, lake, or slow stream; either resting with head downwards, or propelling itself with its powerful hind legs. The sexes of this and other species are so distinct that in the earlier days of entomology they were looked upon as different species. As the habits of this Beetle are almost identical with those of all its family, it will be described at greater length than can be afforded to the generality of insects.

If one of these Beetles be examined on the under side, the metasternum will be seen to have a forked and rather sharp projection from its centre, the points being directed to the end of the abdomen. What may be the object of this curious appendage is not easy to say; certain, however, it is, that it can be used as a weapon on some occasions.

When, for example, an unpractised entomologist catches one of these Beetles in his hand, and has taken care to keep his fingers out of the way of its jaws, he finds himself suddenly and smartly wounded as the Beetle struggles to regain its liberty. The fact is, the insect, led by some strange and unaccountable instinct, always retrogrades when seized in the hand, and so inflicts a rather unpleasant wound with the ends of this appendage. Whether or not it knows of the presence of the weapon, and the use to which it is put, is of course impossible to say; but that the insect can use its forked dagger as well as if it were thoroughly acquainted with it, any of my readers can easily test for himself by going to the nearest pond and catching a Dyticus. Other Water Beetles possess the forked appendage; but it takes different shapes in different species, and is exceedingly useful to entomologists, by enabling them to decide upon the species when other marks fail them.

As the two sexes are so dissimilar in appearance, it will be necessary to describe them separately. The colour of the male Beetle is dark olive, with the margins of the elytra marked with a yellowish streak, narrow towards the apex, and widening considerably towards and on the shoulders. It is in consequence of this streak that the Beetle has received the specific name of *marginalis*. The elytra are very smooth, with the exception of three rows of punctures on the disc. There is a reddish-yellow triangular mark on the forehead, and a very slight ridge on the crown. The thorax is blacker than the elytra, and, like them, has the margin yellow.

The legs of this Beetle are excellent examples of these limbs as they are modified in the Hydradephaga. Both the middle and hind pairs of legs are flattened, oar-like, and furnished with the bristle blade, and the coxa is so made that it only allows one kind of movement to the limb. In consequence of this peculiarity the Dyticus cannot walk properly, but only scrambles about; and if it should by chance fall on its back on a smooth surface, it spins round and round in a most ludicrous fashion.

The first pair of legs, however, are the most interesting. We have already seen that, in very many Beetles, the tarsi of the front pair of legs are dilated in the male, but there are none which even approach those of the Dyticus in complexity of structure. The geodephagous males have the under surface of these dilated joints merely furnished with a pad, but the Dyticus has a most wonderful array of suckers, exactly analogous in principle to those which stud the arms of the cuttle-fish. One of the legs is here shown. The three basal joints of the tarsus are enormously swollen, so that they assume a plate-like shape. Their upper surface is smooth enough, but the under surface is covered with suckers, one of them very large, and the second about half its size, and a multitude of smaller suckers. The larger suckers are placed directly upon the joint,

Dyticus, anterior leg of male.

and the others are at the end of slender footstalks, looking something like the 'patera' champagne glasses with the stems much attenuated.

The female is, as has been stated, very different from the male in appearance. She does not possess the wide tarsi which are so conspicuous in her mate. The elytra are very different from those of the male, and, instead of being smooth and polished, are deeply grooved nearly as far as the middle.

The voracity of this Beetle is really astonishing. It will eat almost any kind of insect, or any kind of meat, raw or cooked, preferring the former. Sometimes it is placed in fresh-water aquaria by those who are not acquainted with its powers, and the result is always disastrous. Not only will it attack every living creature in the aquarium, but during the night it is sure to take to its wings and fly off in search of more food. Even gold fish have been attacked by this insatiate devourer, which does not even spare its own kind, and devours the opposite sex as well as its own. When these Beetles take flight from the pond or stream in which they have been living, they always crawl up a reed or a water-plant, so as to gain space to spread their beautiful wings. In consequence of this freedom of locomotion, this Beetle may be found in almost any pond, however small it may be. Even when ponds have been reduced to mere puddles by the drought of summer, the Dyticus may be found plentifully, hiding itself in the still soft mud.

The mode in which this insect breathes is really wonderful. Being an insect, it is forced to breathe atmospheric air, and yet it has to pass the greater part of its time under water. The problem is solved by the Beetle converting itself for the nonce into a diving bell, renewing the supply of air as often as needful. This feat is accomplished in the following way:—The body is rather flat, so that there is a space between the folded wings and the elytra. Now these elytra are very large, and, when closed, are quite air-tight. When the Beetle wishes to breathe, it comes to the surface of the water with its head downwards, and just exposes the tip of its abdomen to the air. In a moment it has expelled the air which has already been used in breathing, and taken in a fresh supply, with which it dives to the bottom. As the spiracles, or mouths of the breathing tubes, open into the space between the elytra and the abdomen, they can take in the air, and pass it through the system. Sometimes, if the observer will approach very quietly, he may see the Beetles floating with their heads downwards, the tips of their tails just above the surface of the water, and their hind legs spread out so as to balance the body in this strange position. All the Dyticidæ employ this curious mode of supplying themselves with air, but it is most conspicuous in the larger species, and is therefore described in connection with this Beetle.

Having now glanced at the history of the perfect Beetle, we will turn to its larval existence.

There is no possibility of evading the fact, that the larva of the Dyticus is ugly. It is very ugly. It is the crocodile of the insect world, lying unseen in its muddy bed, and darting out at any luckless insect that may pass near it.

When full-grown, this larva is two inches in length. Its colour is yellowish-brown, sometimes one tint predominating, and sometimes the other. The reader will see how this sombreness of hue enables it to lie concealed upon the mud as it waits for prey. At the end of its body are two slender appendages fringed with hairs. These appendages communicate with the breathing tubes which pervade the body, and the larva may be observed in a position resembling that which is assumed by the perfect insect, the head downwards, and the extremity of the tail just above the surface of the water, suspended and balanced by the appendages.

The mode in which this formidable creature obtains its nourishment is very remarkable. The mandibles are large, sharp, and curved. When submitted to a good magnifier, they are seen to be constructed on the same principle as the fangs of a poisonous serpent, a hollow groove running throughout their length. This groove is not left open, but is closed for the greater part of its length by a membrane, an aperture being left at the base. This singular structure enables the larva first to plunge its mandibles deeply into the body of its prey, and then to suck out its juices through the hollow jaws.

As is the case with the carnivorous Beetles generally, the larva soon attains its full growth, and, when the time is at hand for its change into the helpless pupal condition, it takes itself to the bank, up which it climbs, and, burrowing into the damp earth, forms for itself a sort of round cell or cocoon, within which it assumes the pupal form. Should the change occur in the summer, the pupa changes into a Beetle in a fortnight or a few days more, according to the warmth of the weather; but if the larva should retire within its cell in the autumn, it remains dormant during the winter, and does not appear until the following spring. As is the case with dark-coloured Beetles generally, the newly-developed insect is very light in colour and soft in texture, not assuming its hard, dark coat of mail until the expiration of some days.

There are altogether six British species of this genus.

We now pass on to another genus, of which twenty-one species are acknowledged to inhabit England, and will take, as our example, *Agabus bipunctatus*, a figure of which is appended hereto.

It has already been mentioned that the Dyticidæ inhabit equally running or still water, and that they may be found indiscriminately in rivers and ponds. Yet, some species prefer the still, and others the running waters, and the latter insects are almost always of brighter colour than the former. Such is the case with the pretty little Beetle which is now before us. The head is yellow in front, and black on the crown, with two rust-red spots, sometimes fused into one, and sometimes so faint as to be scarcely visible. The thorax is yellow, with the exception of two round black spots on the disc. These sometimes are fused to-

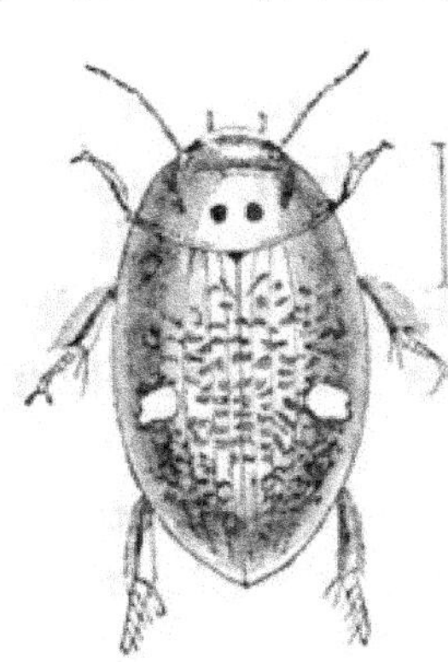

Agabus biguttatus.

gether, like those of the head. The elytra are pale yellow, diversified with small black-brown streaks, a brighter yellow spot near the middle, and a stripe of the same colour down the suture, and upon the lateral margins.

PASSING over several genera we come to a Beetle which is called *Hydroporus duodecim-pustulatus*, and which is selected as an example of a very large genus containing between forty and fifty species. All the Beetles of this genus have their bodies much flattened, and the tarsi of the first and middle pair of legs with four joints, while those of the hind pair have five joints. The two first joints of the antennæ are rather longer than the others.

They are all small Beetles, and the present species is perhaps the largest of its fellows, though it is, on an average, only a quarter of an inch in length. The colour of the head is dull reddish-brown, lighter and redder in the middle. The anterior margin of the thorax has a black stripe in the middle, while a similar stripe on the posterior margin is widened so as to form two black lobe-like marks on the thorax. The elytra are black, and on each elytron are six spots of the same colour as the middle of the head, three of the spots running parallel with the suture, and the other three being placed along the margin. Beneath, the body is yellowish.

This pretty little Beetle is exceedingly common in some places, and correspondingly rare in others. Although the spots differ much in size and shape, and in some specimens are even fused into each other, there is no difficulty in recognising the insect.

OUR next example of the Hydradephaga is *Haliplus variegatus*, an insect which is shown in the accompanying illustration.

Like the last species, this is a pretty little Beetle, and exceedingly variable in its colour, so variable indeed that it has been described by the same writer under the name of at least two species. It is a very small insect, not quite one-sixth of an inch in length. Its usual colour is as follows:—The head is dark brick-red, deepening into blackish-brown on the top. The thorax is paler than the head. The elytra are rather convex, sharply pointed, deep reddish-brown in colour, and have some blackish spots near the margin. This variety is common; but there is one which is much rarer, and in this the general hue is greyish-yellow, and the whole insect altogether lighter in colour. The brightest-coloured specimens are found in rather swift streams running through a gravelly soil.

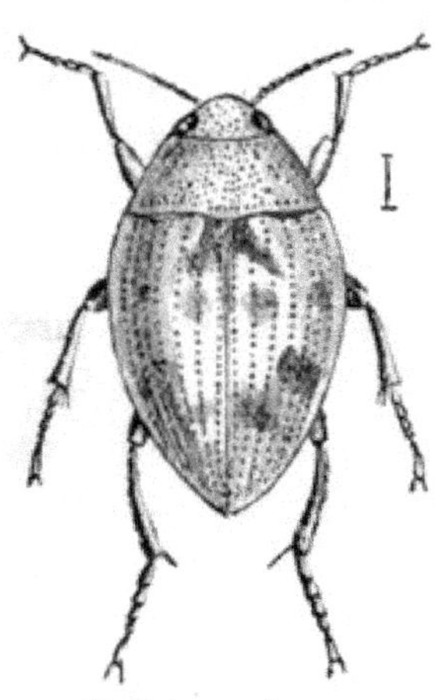

Haliplus variegatus.

There are twelve British species belonging to this genus.

HAVING now gone through the Dyticidæ, we come to the second family of the Hydradephaga—

that of the Gyrinidæ, or Whirligig Beetles, so called on account of the manner in which they whirl themselves about on the surface of the water.

There is not the least difficulty in determining whether or not a Water Beetle belongs to the Gyrinidæ. Besides the distinctions before mentioned, all the Gyrinidæ appear to differ not only from the Dyticidæ, but from Beetles generally, in having, or rather in appearing to have, four compound eyes instead of two. The real fact, however, is that, in order to suit the peculiar habits of the insects, the structure of the eye is modified.

These Beetles pass the greater part of their time on the surface of the water, rowing themselves about with wonderful velocity, and always on the look-out for prey. If the eyes were formed like those of the Dyticidæ, the water would certainly impinge against them and render the insect incapable of seeing by reason of the drops of water which would be continually splashed over its eyes. In order to enable it to see properly above the surface, it is needful that the eyes should be placed high enough to be out of reach of the water; and to enable it to see objects in the water, it is necessary that the eyes should be submerged.

The eyes of the Whirligig Beetles are in fact water-telescopes. Instead of being placed in two masses, one on each side of the head, each is divided by the portion of the head which carries the antennæ; so that half of the eye-cluster is well out of the water and can see objects above the surface, while the other half is submerged, and can see objects beneath it.

The COMMON WHIRLIGIG, *Gyrinus natator*, is very gregarious in its habits, and may generally be seen in small companies, whirling about on the surface of the water in very still and sheltered places. As they dart about, they often strike against each other; but the shock does no harm to their hard and polished bodies, and they go on with their unceasing round as if nothing had happened. Their chief object in thus continually darting over the surface is to obtain food, which consists principally of small flies, Beetles, and other insects which fall into the water. They use their long fore legs in the capture of prey. They are watchful little Beetles, and if they fear danger they dive to the bottom, and there remain until they think that they can return in safety to the surface.

The life history of the Gyrinus is rather a curious one, and is much the same with all the species. The eggs are deposited on a water-plant, and laid in regular rows. From them, in a week or a little more, the curious larvæ are hatched. The larva is dirty-white in colour, and has a large, flat, oval head, armed with powerful jaws, and six rather long legs; while from each side of the eight last joints of the body proceeds a very slender filament, which is part of the respiratory system. The last segment has two pairs of these filaments, each of which is seen, on being viewed by the aid of the microscope, to contain an air-tube, which passes into the body and there joins the general system. When in the water its appearance is very much like that of a centipede, except that the respiratory filaments have no independent motion, like the legs of the centipede, but trail loosely in the water.

In due time the larva is full-fed, and it then, as do many other aquatic creatures, leaves the water and crawls up the stem of a water-plant, until it is several inches above the surface. Having found a safe place, it spins for itself a small grey cocoon, and there waits until it has assumed its perfect state, when it breaks through the walls of the cover, and again seeks the water.

There are ten English species of Gyrinus, some of which are rarer than others. The present species, which is the most common, is about a quarter of an inch in length, and blue-black in colour, with a reddish mouth. The elytra are greenish at the margins, and become narrowed towards the apex. They are very slightly striated and punctured.

CHAPTER V.

BRACHELYTRA.

THE group of Beetles which comes next in order is equally conspicuous with the Hydradephaga, but utterly unlike it or any of the groups which have been described. These Beetles are long-bodied, agile, and seem to play the same part among Coleoptera as the weasel tribe among the Mammalia. Most, though not all of them, are predatorial, and some of them, especially the larger species, are exceedingly fierce as well as voracious, and will fight any foe, no matter how much they may be overmatched.

The name Brachelytra is a very appropriate one, signifying short elytra. These insects have the elytra very short and squared, so short indeed that six or seven segments of the abdomen generally protrude beyond them. Although the elytra are so small, the wings are very large ; and, though they must necessarily be folded in a most complicated manner before they can be packed under the elytra, these insects can take the air with more readiness than any other Beetles, except, perhaps, the Tiger Beetles, whose manner of flight has been before described. In fold-ing the wings under the elytra, the Beetle is obliged to act in a very curious manner, bending the tail over the back, and with the extremity of the body arranging the wings under their sheaths. The earwig uses its forceps for a similar purpose, as we shall see when we come to that insect. The accompanying woodcut shows one of the large Brachelytra in the act of packing up its wings. In consequence of their activity both on the wing and on foot, these insects have gained the popular name of ROVE BEETLES.

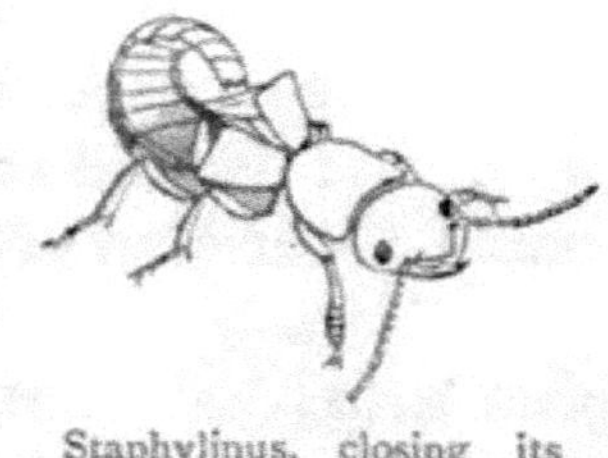

Staphylinus, closing its wings with extremity of its tail.

All these Beetles have the habit of bending their bodies upwards when alarmed, for which reason they have received the popular name of Cocktail Beetles. This act has, in the larger species, so menacing an aspect that many persons are afraid to touch so formidable an insect. In reality, the smaller species are more to be dreaded than the larger. I have already mentioned that the Brachelytra take easily to wing, when they may be mistaken for flies, so ample are their wings and so quick their movements. Many of them are very small—not thicker than an ordinary horsehair—and these are almost invariably the little black 'flies' that are in the habit of getting into the eye on fine summer evenings, and causing an amount of pain which seems quite disproportionate to the size of the insect. Of course even a small fly would cause pain if it got into the eye; but when one of these Beetles finds itself imprisoned, it instinctively turns up its pointed tail, and thus causes a double amount of irritation. I believe that, out of every ten 'flies' that get into the eye, seven are Brachelytra.

Although the larger Brachelytra need not be particularly dreaded, in spite of their fierce looks, it is as well not to handle them without necessity. Their bite, although sharp, is of no particular consequence ; but they possess a more formidable weapon than their jaws. At the end of the tail are two tubercles, which exude a secretion of the most odious character. Like that of the skunk, it has an odour, or rather a stench, peculiarly—and fortunately so—its own, and which cannot be described by any comparison. That of the common snake, when irritated, comes, perhaps, nearer it than any other ; but even that singularly unpleasant emanation is not so utterly disgusting as the effluvium of an angry Rove Beetle.

THE first family is that of the Aleocharidæ. In this family the front tarsi of the males are not wider than those of the females, but the sexes can be distinguished by looking at the last segment but one of the abdomen, and seeing whether it is tubercled, ridged, or has a thicker posterior margin. In such cases the insect is of the male sex.

THE typical genus *Aleochara* has the head deeply sunk into the thorax, which is convex and broad. The elytra are broader than they are long. The abdomen has a flattened margin along the sides, and the tarsi have five joints, the basal joint of the hinder tarsi being longest. The antennæ are short, stout, and the fourth and tenth joints are of equal length.

In the accompanying woodcut is represented a good example of this genus, *Aleochara fuscipes.*

The colour of this Beetle is shining-black. The antennæ are short and thickened in the middle. The elytra are red, edged with black, and the legs and base of the antennæ are red, the thighs being dull-brown instead of red. This insect flies rapidly. It is a common species, and, small as it is, yet is the largest of its genus. It haunts decaying substances, whether animal or vegetable. Fifteen British species are known.

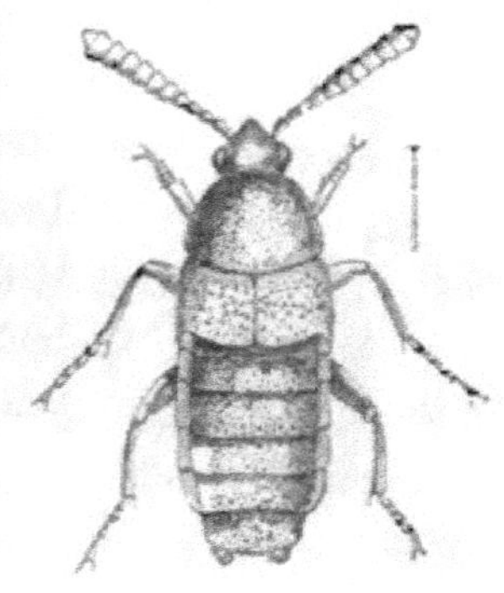

Aleochara fuscipes.

THE Beetle which is our last example of this family belongs to an enormous genus, containing at least 160 species. Its name is *Homalota brunnea*, and it is depicted on page 64.

In this genus the head is without any distinct neck, and the body is narrow and much flattened. The tarsi of the front legs have four joints, and those of the hind legs five joints, the four first joints being equal in size. The joints of the antennæ are bead-like.

The present species is a flat, shining, brick-red insect, with the exception of the head and the last segment but one of the abdomen, which are grey-black, the abdomen being thickly and rather deeply punctured. There is a very shallow groove in the middle of the thorax. The legs are pale reddish-brown.

This is a very common insect, and yet Mr. E. A. Smith, who has long given much attention to the Brachelytra, tells me that he cannot fix upon any special locality for it, having found it indifferently in sand-pits, on palings, and similar places. Indeed, the whole family is a very bewildering and troublesome one to the investigator, and would require the uninterrupted labour of several years before it could be thoroughly mastered.

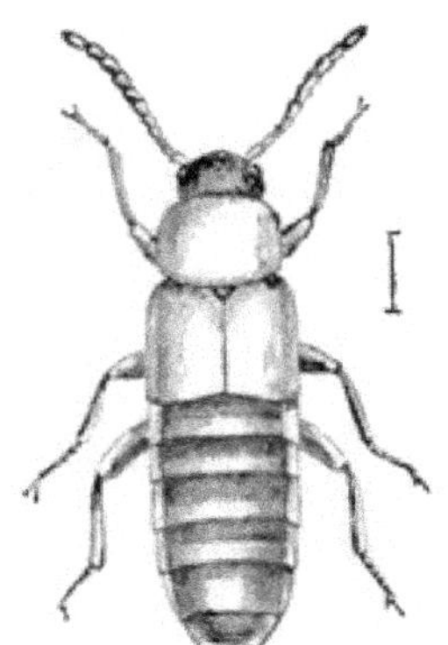

Homalota brunnea.

THE family of the Tachyporidæ comes next in order. These Beetles have the head usually sunk deeply in the thorax, without any distinct neck. The spiracles of the prothorax are conspicuous, and the antennæ are before the eyes, on the margin of the forehead. The males have the basal joints of the tarsi dilated. All these Beetles are unrivalled for their speed, and in consequence of this characteristic the name of *Tachyporidæ*, or 'swift-footed,' has been given to them.

OUR first example of this family is a Beetle called *Boletobius atricapillus*. The insects of this genus live in fungi of different kinds, on which account they are called by the name of *Boletobius*, or 'fungus-inhabiting.' Others are seen as they appear when running in and out of the gills of a mushroom. In this genus, which contains four species, the body narrows to a point behind, the head is long, the palpi slender, and the antennæ are long in proportion to the size of the insect.

The general colour of this species is glossy-red and shining. The head, breast, scutellum, and tip of the abdomen are black, and the elytra are blue-black with a cream-white curved mark on the shoulder, and a line of the same colour upon the posterior margin. The antennæ are rather curiously coloured, the four first joints being black, the next five pale-red, and the last black, like those of the base.

It is a very common insect, and may be found in fungi in the autumn. Indeed, in consequence of their fungus-loving habits, all these Beetles are to be found towards the close of the year. None of the Tachyporidæ are large, and though most of them frequent fungi, many are found under leaf-heaps, in bones, and similar localities.

THE family of the Staphylinidæ, which comes next in order, contains the largest species of this group of Beetles, some of them reaching, or even slightly exceeding, an inch in length. They may be known by several peculiarities of structure. The antennæ are set far apart, their junction with the head being in front, within the base of the mandibles, which are large and formidable. The maxillary palpi are slender, and the ligula small. The spiracles of the prothorax are large. The tarsi of the front feet are dilated in the males and slender in the females. The jaws, too, are not so powerful in the female, neither are their heads so large as is the case with the other sex.

OUR first example of the Staphylinidæ is one of the finest—in my opinion the very finest—of that family. It is called scientifically *Creophilus maxillosus*, but has, unfortunately, no popular name, probably because it is confounded in the popular mind with the common black species, which will be presently described. Its name is more appropriate and expressive than is generally the case with insect names. The word *Creophilus* is of Greek origin, and signifies 'flesh-lover,' while the specific title of *maxillosus* signifies 'large-jawed.' Both names show that those who affixed them to the insect were thoroughly acquainted with its character and form, for the Beetle is a most voracious carrion eater, and has jaws that are of enormous size in proportion to its body. The colour of this Beetle is shining-black, but it is mottled with short grey down.

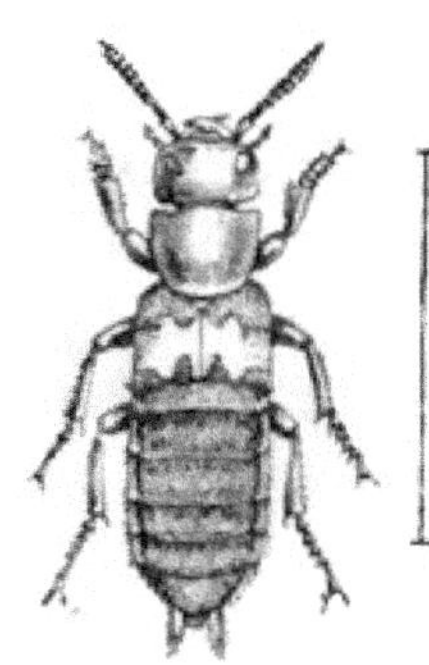
Creophilus maxillosus.

In some places this Beetle is tolerably plentiful, but in others it is seldom if ever seen. It can generally be captured in the bodies of moles that have been suspended by the professional mole-catchers—and, indeed, these unfortunate moles are absolute treasure-houses for the coleopterist, as we shall see when we come to the next group of Beetles. It is the only British species of its genus, and can be distinguished by having short and thickened antennæ, smooth head and thorax, and the latter rounded.

NOW comes an insect that is very familiar to us,

the DEVIL'S COACH-HORSE, as it is popularly and fancifully called. Its scientific name is *Ocypus olens.*

I really think that this is the very ugliest insect in England. It is scarcely so repulsive as the cockroach, its wonderful courage and spirit being redeeming points, but it is so very hideous, that the popular name exactly expresses its appearance. Its colour is dull, dead-black; its eyes, which scarcely project from the head, have a cold, cruel look about them, and its tail, when raised menacingly like that of a scorpion, protrudes two yellow vesicles at the tip, from which emanates the horrible odour that has been already mentioned. Sometimes it finds its way into cellars and larders, if they be wholly or partially underground; and then the servants are always much alarmed at the creature, of which they have a dread which is superstitious rather than the offspring of mere prejudice or ignorance.

This is one of the most active of Beetles. Being furnished, like all its family, with long and ample wings, and not being burdened, like the chafers, with a thick and heavy body, it flies with great rapidity, and can pass over incredible distances without being obliged to rest. It is equally active on the ground, for which reason the generic name of *Ocypus*, or 'swift-footed,' has been given to it. Some writers on entomology have given to this insect the generic name of *Goerius*, or 'mournful,' in consequence of its sombre and funereal colouring. The specific title *olens*, or 'stinking,' is given to it in consequence of the horrible emanations from the tail tubercles.

We will now glance at the life history of this

Beetle, which, in spite of its ugliness, is really a very interesting one.

The eggs of all the Staphylinidæ are large in proportion to the creature which produces them, but those of the Devil's Coach-horse are larger than those of the largest British insect, being one-tenth of an inch in length and one-twelfth broad.

When these are hatched, little larvæ issue from them, somewhat similar in form to the parent insect, though, of course, without any vestige of wings. These larvæ are quite as fierce as the perfect insects, and much more voracious, eating being indeed, as with all larvæ, the chief business of their lives. They are predacious, and, though they will devour carrion when they can procure it, will attack and kill any insect which comes near them, not even sparing their own kind. They have an ingenious mode of seizing their prey in the soft interval between the head and neck, and then, plunging their sharp and curved jaws deeply into its body, they suck out its juices.

They can be found throughout the spring, and may often be captured by digging shallow holes in the ground in some sheltered spot, placing a piece of meat, a dead bird or a frog, in the hole, and covering it with a stone so as to protect it from the elements, but leaving space for the ingress and egress of the Beetles. Towards the end of spring or the beginning of summer, the larva is full fed, and burrows a hole in the earth, in which it undergoes the change to the perfect form.

There is a strange peculiarity about the pupæ of these Beetles. With nearly all wing-bearing Coleoptera, the wings are folded under the elytra, even though the latter organs be comparatively short; but, in the pupæ of these curious Beetles, the wings are extended beyond the elytra and fold over the breast, so that two-thirds of their length is seen beyond the elytra. They remain in the pupal state for a fortnight or three weeks, and then assume the perfect form. The Beetle is most plentiful in the autumn. I strongly recommend any of my readers not to injure this Beetle, repulsive as it may appear. It does no harm, either to the garden, the orchard, or the field, but, on the contrary, from its inveterate insect-eating habits, rather confers a benefit on the agriculturist.

There are ten British species of this genus, which is known by the long thread-like antennæ, with the last joint oblique at the tip, and the large head.

Philonthus marginatus.

OUR next example of the Staphylinidæ is an insect of much less size and very different shape, called *Philonthus marginatus*, the only specimen that we can take of the very large genus, of which forty-seven British species are acknowledged. In this genus the thorax is squared, the antennæ and palpi are slender, and there is a strong tooth in the middle of each mandible. The name *Philonthus* signifies 'dung-loving,' and is given to this genus because the largest and most conspicuous species are found under patches of cow or horsedung. Some of the smaller species, however, live under heaps of decaying sea-weed, such as *Philonthus fucicola*, the latter term signifying some creature that inhabits sea-weed. The present species is generally to be found under dead leaves.

The colour of this little Beetle is black, but the legs and the margin of the thorax are reddish-yellow, a peculiarity from which it derives its specific name *marginatus*. The middle segments of the abdomen have their edges yellow. This is a very curious Beetle.

NEXT comes the family of the Stenidæ. These insects may be known by the position of the antennæ, which are generally set between the eyes or on the front margin of the forehead. The basal joint of the maxillary palpi is long, and the last joint almost imperceptible.

Sometimes the young entomologist is much puzzled by a phenomenon which takes place with sundry small Beetles belonging to the Brachelytra. As soon as they are killed a long and slender tongue-like organ darts from the mouth, and protrudes itself until it looks like a proboscis. These Beetles belong to the genus Stenus, of which we have an example in *Stenus bimaculatus*, which is herewith shown. This tongue-like organ is in fact composed of the ligula, the two paraglossæ, and the labial palpi. The mandibles of this

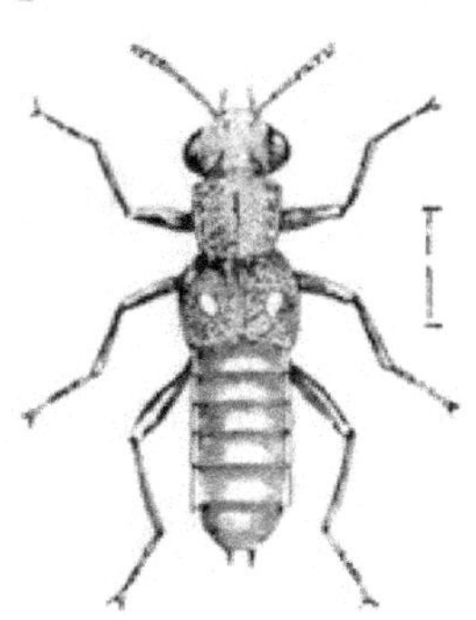

Stenus bimaculatus.

genus are strongly formed, having one very large tooth, and four very small teeth. The fourth joint of the tarsi has a slightly double lobe.

The present species is black, with the exception of a round tawny spot on each elytron, from which the insect has derived its specific name of *bimaculatus*, or 'two-spotted.' The surface is thickly and deeply punctured, and is covered with a scanty whitish down. There is a deep furrow along the middle of the thorax. The legs are tawny, with the exception of the knees and tarsi, which are black.

This is a very good example of the genus, and indeed so well exhibits the characteristics of the family that it is worth a detailed examination. It is a very common insect, being found all over England, and almost always to be taken on the banks of ponds and rivers among the aquatic plants. In tolerably warm weather it may be taken running about upon the stems of the plants, and on cold days lurking in the muddy soil about their roots. Mr. E. A. Smith, to whom I am indebted for much information concerning the smaller Brachelytra, tells me that all the spotted Steni are found in the wettest situations. This genus is a very large one, containing more than fifty species.

THE next family on our list is that of the Homalidæ. In these Beetles the spiracles of the prothorax are hidden, the antennæ are set on the sides of the forehead, and the maxillæ have a horny hook at the tip. There are two ocelli, or simple eyes, upon the back of the head, this being a very valuable characteristic in arranging these insects. As a rule the Homalidæ have flattened bodies, and long slender antennæ, and delight in damp places, whether wetted by fresh or salt water; so that they can be found under heaps of decaying sea-weeds on the coasts, and under stones on the banks of ponds. Some of them may be found under bark, and some in flowers; so that they have a very wide range of locality.

OUR typical example of the family is shown in the accompanying woodcut, its name being *Homalium florale.* In this genus the body is rather oval and flattened. The antennæ are short and hairy, becoming thicker at the tips. The thorax is short, somewhat heart-shaped, and narrowed behind. The four basal joints of the tarsi are short.

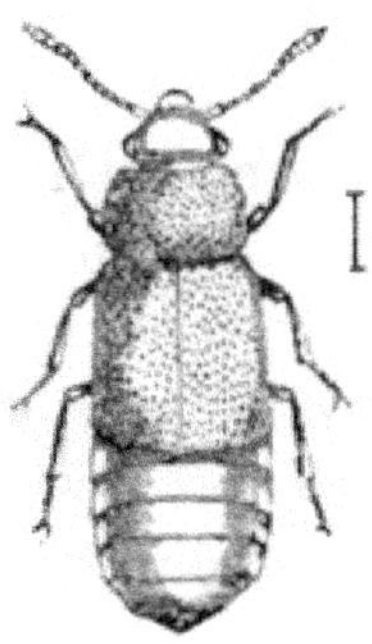

Homalium florale.

The species which serves as our example of the typical genus is greyish-black and shining, the surface being thickly punctured, and the punctures inclined to form striæ on the elytra. The legs are reddish, and the antennæ and palpi black. This is one of the flower-loving species, being found in the spring time frequenting the flowers of the hawthorn and sallows. It is distributed over England generally, but does not seem to be plentiful in any particular locality. Twenty British species of this genus are known.

CHAPTER VI.

NECROPHAGA.

THE reader will probably observe that, in the groups of insects which have already been described, allusion has been made to the analogies between them and certain groups of vertebrates. The Geodephaga, for example, represent the land Carnivora, the Hydradephaga those of the water, and the Brachelytra represent in some degree the slender and lithe-bodied weasels. We now come to a group which takes among insects the part which is played among the higher animals by the hyenas and vultures, these Beetles being the scavengers of the insect world.

The name NECROPHAGA, i.e. 'carrion-eaters, which distinguishes this group, is expressive of their character. By some systematic entomologists they are called CLAVICORNES, or 'club-horned,' because their antennæ, slender at the base, are expanded at their tips into a rounded knob. This shape of antennæ is called 'clavate' by entomologists. The form of the antennæ is an admirable characteristic by which these Beetles may be known. There is no palpus on the inner lobe of the maxilla, the scutellum is always conspicuous, and the elytra are wide, though not always long. Indeed, in many species they do not nearly reach the end of the abdomen, and are quite as short as those of many Brachelytra. In such cases they are generally 'truncate,' i.e. looking as if they had been cut off square.

THIS is the case with the first family of Necrophaga, the Silphidæ. The mandibles are powerful, as is required for the work which they have to perform, there is a very distinct labrum, and the trochanters of the hind legs are projecting. These are again subdivided into two sub-families; the first being called Silphina, and the latter Cholevina. The former sub-family may be known by the fact that the antennæ have ten joints, and a very distinct and rounded club. Their wings are very large and powerful, as is needful for insects whose food is necessarily scattered over a very wide area. It is worthy of notice that, when they are flying, their elytra are carried very upright, so that their backs approach quite closely to each other.

The first genus of the Silphidæ is Necrophorus, a word which signifies 'carrion-bearer,' in allusion to the singular habits possessed by all the Beetles of this genus. They do not content themselves with merely eating their food, but they bury it, and then lay their eggs in it, so that it serves not only as a feast for themselves, but as a provision for their future young. In consequence of this habit, they go by the popular name of BURYING, or SEXTON BEETLES. It is a very appropriate name, for there is scarcely any dead animal or portion of an animal which they will not contrive to bury; and if it be too large for one Beetle, several others will take a share in the work.

They will bury birds, frogs, rabbits, pieces of meat, or anything of a similar kind, and do it with wonderful rapidity; thus rendering a doubly important service, by removing the decaying animal matter from the surface of the earth, and helping to fertilise the ground by burying it below the surface. The manner in which these Beetles execute so difficult a task is admirably told by Mr. E. Newman, in his 'Letters of Rusticus:'—

'Two days after, I was again in Godbold's; and seeing the bullfinch lie where he had been left, I lifted him up by the leg, intending to make a present of him to a fine colony of ants established, a little further on, in the days of General Oglethorpe, and which had maintained their station ever since. They had made many a pretty skeleton for me, and I intended to add that of a bullfinch to the store; but the buzz of a Beetle round my head caught my ear. He flew smack against the bullfinch, which I was holding up by the leg, and fell at my feet. I knew that the gentleman was a Burying Beetle; and as I put the bird down for him, he soon found it, mounted upon it, and, after much examination, opened out his wing-cases and flew away. I will profit by his absence to tell you a bit of his history.

'The Burying Beetle is about an inch in length; he is black, with two bands across his back of a bright-orange colour: these bands are formed by two blotches of that colour on each of the wing-cases. He is a disgusting creature though in such a gay dress, being so fetid that one's hands smell for hours after handling him; and if he crawls on one's coat, or other garments not often washed, the smell continues for days. The whole tribe of Burying Beetles lay their eggs in the bodies of dead animals, which, when possible, they bury for the purpose.

'The Burying Beetles hunt in couples, male and female, and when six or eight are found in a large animal, they are almost sure to be males and females in equal numbers. They hunt by scent only, the chase being mostly performed when no other sense would be very available—viz. in the night. When they have found a bird, great comfort is expressed by the male, who wheels round and round above it like an eagle; the female settles on it at once, without this testimonial of satisfaction. The male at last settles also, and the bird undergoes the scrutiny of four at least of the senses—touch, smell, sight, and taste—for their heads are continually diving among the feathers of the bird, and a savoury and ample meal is made before the great work is begun. After the Beetles have appeased the calls of hunger, the bird is abandoned for a while; they both leave it to explore the earth in the neighbourhood, and ascertain whether the place is suitable for interment. If on a ploughed field, there is no difficulty; but if on grass or among stones, much labour is required to draw the body to a more suitable place.

'The operation of burying is performed almost entirely by the male Beetle, the female mostly hiding herself in the body of the bird about to be buried, or sitting quietly upon it, and allowing herself to be buried with it. The male begins by digging a furrow all round the bird, at a distance of about half an inch, turning the earth outside. His head is the only tool used in this operation; it is held sloping outwards, and is exceedingly powerful. After the first furrow is completed, another is made within it, and the earth is thrown into the first furrow; then a third furrow is made, which being under the bird, the Beetle is out of sight. Now the operation can only be traced by the heaving of the earth, which soon forms a little rampart round the bird; as the earth is moved from beneath, and the surrounding rampart increases in height, the bird sinks. After incessant labour for about three hours the Beetle emerges, crawls upon the bird, and takes a survey of his work. If the female is on the bird, she is driven away by the male, who does not choose to be intruded on during the important business.

'The male Beetle then remains for about an hour perfectly still, does not stir hand or foot; he then dismounts, diving again into the grave, and pulls the bird down by the feathers for half an hour. Its own weight appears to sink it but very little. The earth then begins heaving and rising all round, as though under the influence of a little earthquake; the feathers of the bird are again pulled, and again the bird descends. At last, after two or three hours' more labour, the Beetle comes up again, gets on the bird, and again takes a survey, and then drops down as though dead, or fallen suddenly fast asleep. When sufficiently rested, he rouses himself, treads the bird firmly into its grave, pulls it by the feathers this way and that way, and, having settled it to his mind, begins to shovel in the earth. This is done in a very short time, by means of its broad head. He goes behind the rampart of earth, and pushes it into the grave with amazing strength and dexterity, the head being bent directly downwards at first, and then the nose elevated with a kind of jerk, which sends the earth forwards. After the grave is thus filled up, the earth is trodden in, and undergoes another keen scrutiny all round, the bird being completely hidden; the Beetle then makes a hole in the still loose earth, and, having buried the bird and his own bride, next buries himself. The female lays her eggs in the carcass of the bird, in number proportioned to its size; and after this operation is over, and the pair have eaten as much of the savoury viand as they please, they make their way out, and fly away in quest of further adventures.'

A VERY common species of this genus is the large *Necrophorus humator*, which may be found in all parts of the country. This is a large, though not brightly-coloured species. It sometimes exceeds an inch in length, and its colour is deep, shining-black, with the three last joints of the antennæ reddish-yellow. The fringe-like pads of the tarsi are of the same colour.

An equally abundant species is *Necrophorus vespillo*, which is very much smaller than the preceding insect. It is very different in appearance from the *Necrophorus humator*, the ground colour being black, while two broad, waved bands cross the elytra, one at the base, and the other towards the apex, both being connected by a narrow band of the same colour upon the margin. The hinder tibiæ are not straight, as in all the other members of the genus, but are strongly curved, like those of a bow-legged man.

We will now trace the progress of the insect from the egg to the perfect Beetle.

Soon after they are deposited, the eggs are hatched ; the larvæ being rather long, fleshy, narrowed at each end, and having the segments, or rings, of the body very distinctly marked. The legs are very tiny, and much too small to move the large, heavy body. A curious substitute for legs is, however, found. On the upper surface of each segment is a horny plate, with strongly-toothed edges. By alternately elongating and shortening its body, the creature is able to force its way through the soft material on which it feeds, just as a snake glides upon the ground, or the worm beneath it. Imperfect as these appliances may seem to be, they enable the larva to scramble along with tolerable speed.

When the larva has attained the length of an inch and a half, and is full-fed, it prepares for its change into the pupal state. This it does by ceasing to feed, and making for itself a sort of cell or cocoon under the ground, in which it casts off its larva skin, and becomes a rather odd-looking pupa, having the end of its tail armed with two sharp spines, by means of which it is able to turn itself about in its cell, from which it emerges, in the spring, a perfect Beetle.

THE genus Silpha is known by the flattened body, the antennæ being less boldly clubbed, and having eleven joints instead of ten. There are fourteen species inhabiting England, and they are, with one exception, much smaller than the members of the preceding genera. The handsomest of the British Silphæ is called scientifically *Silpha thoracica*, and can be identified at a glance. This fine insect is readily known by its colouring. The head is black, and the thorax is brick-red, covered with a very short golden down, and much crumpled. The elytra are very much like those of the preceding insect, being black, and traversed longitudinally by three ridges, the second and third of which are connected by a raised tubercle.

The larvæ of all the Silphas are very different from those of Necrophorus. Instead of being sluggish, fat, long-bodied grubs, they are active, flat, and wide, running about with wonderful velocity. A heap of old marrow bones is a very favourite haunt of these larvæ, and, if the bones are tapped so as to disturb without hurting their inmates, it is wonderful to see how the flat, black larvæ come scurrying out, looking very much like black wood-lice, and perfectly well able to take care of themselves; while the larvæ of Necrophorus lie utterly helpless on the surface of the ground. Silphæ may be found in much the same localities as the preceding insects. The best places to find them are, however, the moles that are too often seen suspended on twigs, and the more moles there are near each other, the richer will be the harvest of Silphæ. I have found that ten moles on one branch contain many more Silphæ than double the number scattered over a wide area.

But the 'happy hunting grounds' of any entomologist who is looking after Necrophaga are the 'keepers' trees,' those monuments of misguided energy. The best keepers' trees, in an entomological point of view, are those of the New Forest, and on them are found the carcases of owls, weasels, stoats, hawks, magpies, ravens, and now and then a rare bird or two, such as the honey-buzzard. The suspended carcases look quiet enough, but when the net is held under one of them, and a series of taps administered, it is wonderful to see how it swarms with animal life. First, out come Beetles of various kinds, some trying to fly away as soon as they reach the open air, but most letting themselves fall into the net. Next comes a whole swarm of larvæ, and, when the bird is a large one, it really seems as if the creatures never would cease from pouring out. Those who wish to collect and watch the habits of these Beetles cannot do better than make a preserve for them by hanging up the body of a rabbit, a puppy, a kitten, or some such creature, so that it may be within reach of the eye, and out of the reach of any except winged devourers.

The word *Silpha* is Greek, and signifies an ill-smelling insect, but the nomenclature of the ancient writers is so uncertain that we cannot absolutely identify the name with the insect. The specific name *thoracica* refers to the very conspicuous colour and the large size of the thorax.

THE sub-family of the Cholevina are known by their narrower bodies, and their heads being sunk in the thorax. One of these Beetles, belonging to the typical genus, is known in science as *Choleva angustata*, but possesses no popular title.

The genus, of which there are seventeen British species, is known by its narrow body, its long and slender legs and antennæ, the very obtuse hinder angles of the thorax, and the oval and striated elytra. The head is black and shining, the thorax is pitchy-black, becoming greyish at the margins and posterior angles. The elytra are rather variable in colour, black being the leading hue, but they take shades of red or grey according to the individual. The apex is round and short, and each elytron is marked with seven faint striæ. The legs are pale reddish-yellow.

The body is covered with a very fine and very short yellowish down. It is not a large insect, seldom exceeding one-sixth of an inch in length.

It is not uncommon to take a specimen that is pale rusty-red or yellow, the reason for this paler hue being that the Beetle has only recently emerged from the pupal state, so that the atmosphere has not exercised its full influence upon it. When it has been exposed for a few days to the air and light, the reddish-yellow will change to the dark-red or grey-black which is the usual colour of the Beetle.

THE important family of the Histeridæ now comes before us. These are flattish square-bodied Beetles, with a sort of steely look about them, and as hard to the touch as if their elytra were veritable pieces of plate armour. It is no easy task to get a fine entomological pin through these creatures, the pin either bending, or its point repeatedly slipping off the hard and polished surface of the Beetle in a manner calculated to injure the temper as well as the pin. I always used to keep by me a rather fine needle fixed in a handle, and projecting about one-third of an inch, so that I could pierce the hard elytra with the steel point, and then introduce the pin. This needle was useful in setting many other hard-bodied Beetles, especially some of the weevils.

These hard and glossy elytra are much shorter than the body, and abruptly truncated ; but below them may be seen a beautiful and wide pair of wings packed away with wonderful neatness. The basal joint of the antennæ is very long, and the club is boldly marked, the three last joints being almost fused into a globular and velvety knob. The generic name of *Hister* is derived from a Latin word signifying an actor or a mimic, and is given to these Beetles on account of their habit of simulating death when alarmed. The popular name of MIMIC BEETLES is often given to them by entomologists.

One of the handsomest species of this genus is the FOUR SPOT MIMIC BEETLE (*Hister quadrimaculatus*). The colour of this insect is black, but on each elytron is a large C-shaped red mark reaching from the base nearly to the apex. This mark is often divided in the middle, so as to produce the effect of four red spots on the back. A narrower line of the same colour mostly runs along the margin ; but this, like in many other Beetles, is apt to be exceedingly variable in the arrangement of its colouring.

This, together with the rest of the genus, can be

found in or under decaying animal matter, patches of cowdung being favourite resorts. The pertinacity with which these insects will feign death when captured is most remarkable, for they will endure almost any amount of rough handling without giving the least signs of life, the legs being folded flatly under the body so that they are scarcely visible. Indeed, I scarcely know whether they or the Pill Beetles, which will presently be described, are the most obstinate in this respect.

The larvæ of the Mimic Beetles are also to be found in cowdung, and are not in the least like their parents, being long-bodied, cylindrical, whitish in colour, with two forked appendages at the tail. Some species prefer dead animals, and may be found in the moles to which reference has already been made. About fourteen British species of Hister are known; and the young entomologist will find that, owing to variations of colour, he will often be rather perplexed to determine the exact species of some new capture.

WITH great reluctance I am obliged to omit several genera, and must proceed at once to the next family, the Nitidulidæ. The Beetles of this family have short clubbed antennæ. The head is sunk in the thorax as far as the eyes, and the mandibles are notched at the tip; the thorax is rather square, and the tarsi have five joints. The body is flattened. The little Beetles which are found in such numbers in flowers, and have such shining bodies, mostly belong to this family, to which the name of *Nitidulidæ* has been given on account of their glittering bodies.

The typical sub-family, the Nitidulina, are known by the length of the elytra, which reach to the last joint of the abdomen, and the shape of the thorax, which does not cover the base of the elytra. One of these insects, *Omosita discoidea,* is a very plentiful species, and may be found in fungi, under the bark of trees, and even in old bones.

The body of this Beetle is oval, and, as may be inferred from its name, is much flattened. Its colour is rusty-brown, and the surface is thickly punctured. Upon the elytra are some faint oblong black streaks, and a curved mark of paler hue than the rest of the body.

In the accompanying woodcut is shown another of these Beetles, *Meligethes æneus.* This genus is known by the squared and highly-polished metallic body, the long elytra, and the third joint of the antennæ, which is exactly as long as the fourth and fifth together. All the Beetles of this genus are very small, and are invariably to be found in flowers, creeping from their hiding places under the petals when the flower is gathered or shaken. Being very minute insects, a careful examination with a somewhat powerful lens is needed to distinguish the species, and, even then, the little creatures are so like each other in size and colour, that the entomologist is obliged to abandon the usual mode of

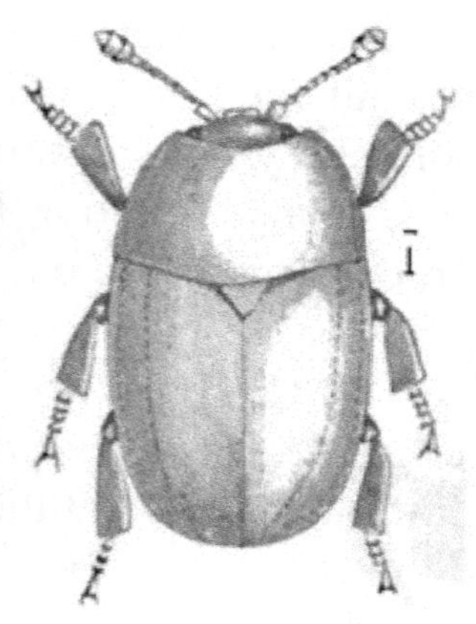

Meligethes æneus.

determining species, and to trust to the number of notches in the tibiæ of the first pair of legs.

This species is variable in colour, being blue-black, violet, or dark green-blue, over which is a sort of brassy gloss. It seldom exceeds the twelfth of an inch in length, and is a very good example of its genus. It is plentiful throughout England.

STILL keeping to the same family, we take another of the sub-families, the Ipsina, which have the front of the head lengthened and covering the labrum, the fourth joint of the tarsus being very minute. Our first example of these insects is *Rhizophagus ferrugineus*, a fairly plentiful species. These Beetles have much narrower bodies than the preceding, the antennæ are short and boldly clubbed, with a large basal joint. The head is not sunk in the thorax, and the elytra are not so long as the abdomen. They are mostly to be found under the bark of trees, but some are fond of inhabiting old bones, and are even parasitic in ants' nests. The name *Rhizophagus* signifies 'root-eating,' and there are about ten British species. Though they are for the most part vegetable-feeders, some at least of the species are known to be carnivorous, and have been detected in eating the larvæ of other bark-feeding Beetles belonging to the genus Hylesinus.

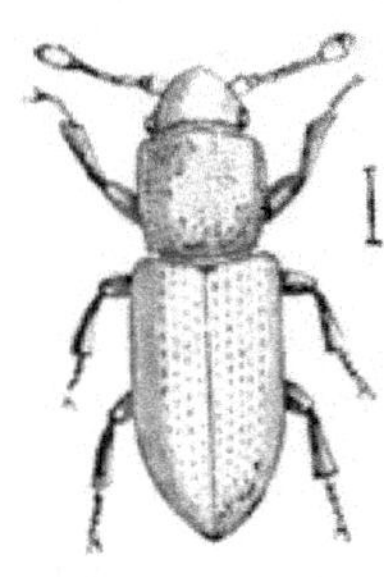
Rhizophagus ferrugineus.

The colour of this species is smooth shining rust-red, sometimes deepening into reddish-black. There is a pit on either side of the head, the elytra are thickly punctured and striated, and beneath it is rust-red. This Beetle has been chosen because it is the largest of the genus, sometimes exceeding one-sixth of an inch in length.

THE family of the Cryptophagidæ will be represented by one example.

One example of this family is *Cryptophagus pilosus*, which is shown in the accompanying illustration. The genus is known by the shape of the margins of the thorax, which are more or less toothed. The present species is oblong, and its colour rust-red, the surface of the body being sparingly covered with very fine down. The thorax is thickly punctured, especially on the disc. In this Beetle the marginal toothing of the thorax is not so conspicuous as in most of the species, and is rather undulated than toothed. It is to be found in and about fungi. There are about twenty-three species of this genus.

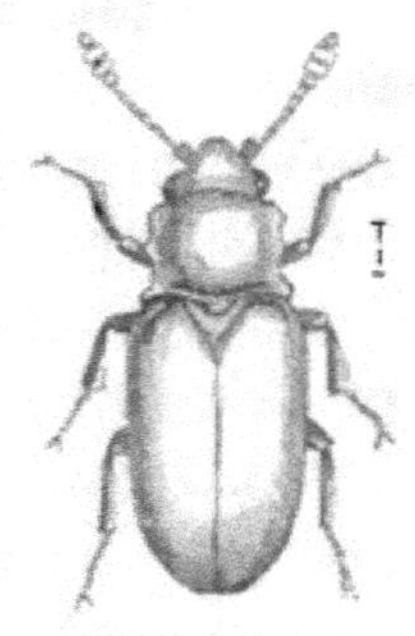

Cryptophagus pilosus.

PASSING by a number of insects, we come to the family of the Dermestidæ or Museum Beetles, a very interesting one to the entomologist, although he can but hold its members in bitterest hatred. The word *Dermestes* is of Greek origin, and signifies 'skin-eater.' The name is but too appropriate, as all possessors of zoological collections know to their cost. It is the Dermestes which forces taxidermists to use the dangerous arsenical soap in their preparations, and it has been the means of depriving many a hard-working man of his best teeth, the arsenic loosening them so that they fall out almost at a touch. Full many a valuable museum has been utterly ruined by these destructive Beetles, which, even when the skin is poisoned with arsenical soap, will attack the hair or the feathers, and strip the creature as bare as if it had been shaved.

This family is distinguished by their short, straight, and doubled antennæ, their small and retractile head, the five-jointed tarsi, and the length of the elytra, which cover the abdomen. The palpi are thread-like and shorter than the maxillæ, and the first joint of the tarsus is shorter than the second. For illustration of this genus I have selected the well-known BACON BEETLE (*Dermestes lardarius*).

This is really a pretty, though not gaily coloured, Beetle, its body being black and its elytra having a very broad greyish band across the base, on which are three black or pitchy spots. On examination with a lens, the band is seen to be composed of a short but thick grey down, the black spots being simply places on which the down does not grow, so that the black of the elytra is rendered visible.

This Beetle may be found plentifully in the 'keepers' trees' which have already been mentioned; and even after the animals have been so dried by exposure that their skins are as hard as horn, the Dermestes will attack them, its sharp teeth enabling it to overcome the hardened skin. The chief havoc caused by this Beetle is due to the larva. Its colour is whitish-brown above and white below, and it is profusely covered with long hairs. The cast skins of these larvæ may be seen abundantly when the Beetle has taken possession of any place, and by them the museum owner is often warned of the danger which has come on his collection. The reader will see that, like many other destructive insects, it is most valuable in its right place, and does good service by removing from sight objects which are not only unpleasant to the eye and nostril but injurious to the health. In these places it should be protected and encouraged; but when it makes its way into a house, extermination is the only course to be used.

WE now come to the Byrrhidæ, or Pill Beetles, so called from their rounded shape, and the manner in which they can hide their limbs and antennæ when alarmed. There is no difficulty in distinguishing Beetles belonging to this family. The antennæ are gradually thickened towards the extremity, and the head is very small and deeply sunk in the thorax, with which it can be completely retracted in most of the species.

The machinery by which the legs are packed up is extraordinary, and this alone would serve to indicate the family. On the tibiæ there is a groove in which the tarsi are received when doubled, the tibiæ fold closely to the femora, and the whole leg, thus reduced into a very small compass, is received into a groove under the body. In fact, the legs are packed up very much like the joints of a portable easel. The head being at the same time withdrawn into the thorax, the antennæ lie pressed closely against its sides, so that when the Beetle has thus packed away all its limbs, it does not bear the least resemblance to an insect. This mode of concealment, or rather of evasion, is rendered more perfect by the fact, that the surface of the body is covered with fine down, which retains the dust of the roads on which it so often travels, and gives to the Beetle the aspect of a little round dusty stone. And, so pertinaciously does it keep this attitude when alarmed, that it will suffer its limbs to be torn from its body rather than give the least sign of life.

The typical genus has the antennæ rather flattened, and shorter than the thorax, the basal joint being large, the second small and globular, and the third long and slender. The club is formed by a series of joints regularly increasing in size, the last joint being egg-shaped. The thorax is waved behind, and the body is very convex.

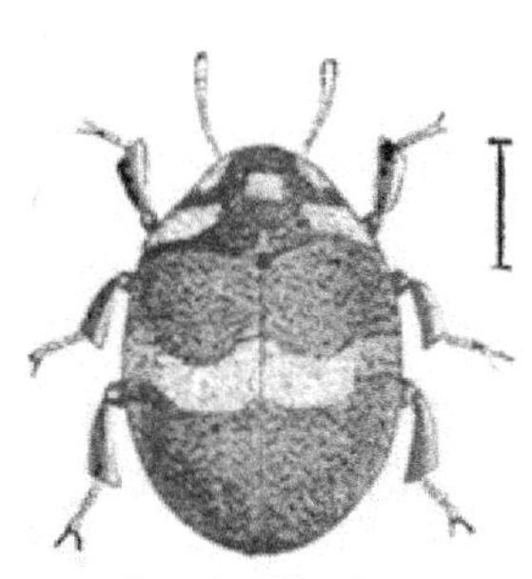

Byrrhus fasciatus.

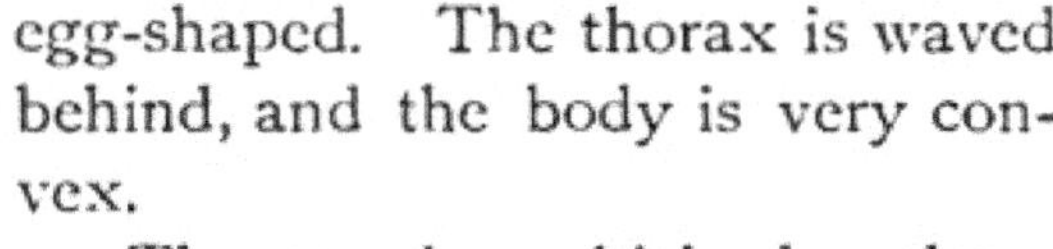

The species which has been chosen for illustration is the BANDED PILL BEETLE (*Byrrhus fasciatus*), a figure of which accompanies this description. The colour of this Beetle is black, the thorax having a decided golden tinge. Upon the elytra are a number of very short black stripes, and in the middle is a reddish-yellow band, shaped as is seen in the figure. This is a tolerably plentiful species,

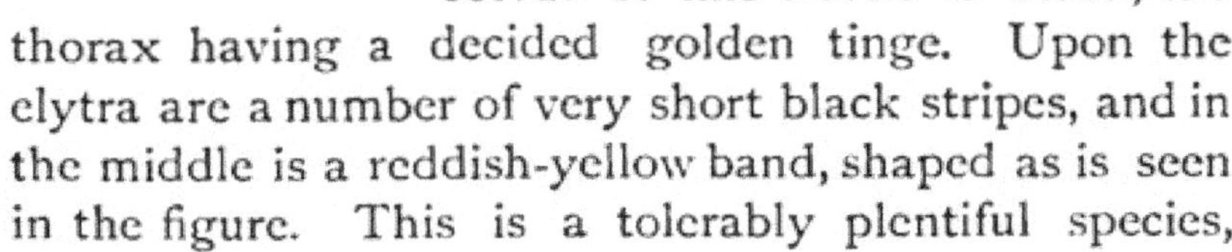

though it is not so often found as the COMMON PILL BEETLE (*Byrrhus pilula*), which is without the yellow band across the elytra. Five species of this genus inhabit England.

NOW we come to a very familiar but little understood insect, popularly called the BLACK WATER BEETLE. Like some of the preceding species, it has been confounded with the Hydradephaga, merely because it inhabits the water, no reference being made to its structure, or even its mode of feeding. Indeed, I believe that scarcely any, except entomologists, have the least idea that the Dyticus and the present Beetle are not the same insect, the only difference being that one is much larger than the other. Now, if we examine this Beetle, *Hydröus piceus*, we shall find a vast number of structural differences, as well as complete divergence in habits.

The Beetle belongs to the family of Hydrophilidæ, i.e. water-lovers. Sometimes the two halves of the word are transposed, the Beetles being called Philhydrida, the signification being exactly the same in both cases. The family may be known by the size of the palpi, which are as long as the antennæ and sometimes longer—the latter organs never having more than nine joints, and sometimes only six—and by the double lobe of the maxilla and the very short mandibles. The tarsi have five joints. The typical genus has the mandibles armed on the inside with three strong teeth, notched at their tips; the second joint of the maxillary palpi is very long, and the elytra become narrower towards the apex.

The insect is herewith represented of its natural size, showing that, with the exception of the Stag Beetle, it is our largest coleopterous insect. A glance at the figure will show one of the principal peculiarities of this Beetle, namely, that the palpi far exceed in length the antennæ, and project in front considerably before those organs. The sexes are easily distinguished by a glance at the antennæ, and the tarsi of the first pair of legs. The metasternum terminates in a sharp spine, so long that its point reaches beyond the coxæ of the last pair of legs.

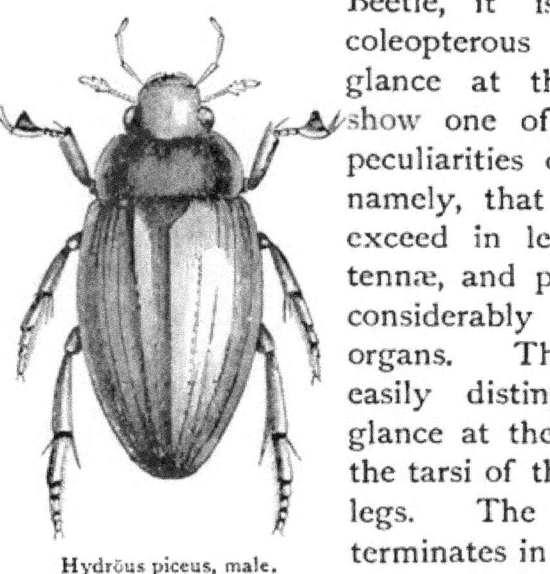

Hydrōus piceus, male.

The colour of this Beetle is smooth blackish-olive, the margins of the elytra taking a bluish tinge. The wings themselves are very large, and have a very fine effect when spread. On each of the elytra are eight striæ, and the breast is clothed with thick yellow down. The metasternum is developed, as in the Dyticus, into a pointed process, but in this insect the weapon is simply needle-shaped, and is about the size of an ordinary darning-needle. It is black, highly polished, and very sharp. There is much variation in tint in different individuals, but the average hues are as given above.

The life history of this insect is a very curious

one. The female Beetle is furnished with a complete silk-spinning apparatus, the spinnerets being placed, not in the mouth, as is the case with the silkworm, but at the end of the tail. With these instruments she forms a cocoon shaped almost exactly like a turnip, being round, and having a pointed projection from one side. Within this cocoon, which soon becomes hard and water-tight, and is fastened to the stem of a water-plant, the eggs are placed; and in a time varying from a fortnight to six weeks, according to the warmth of the weather, the larvæ are hatched. About fifty or sixty eggs are placed in one cocoon, and by this extraordinary provision they are preserved from harm until the larvæ are able to escape into the water and shift for themselves. They are very small at first, but rapidly increase in size until they attain the length of three inches.

The larva is soft, fat, and dusky in colour, and feeds voraciously on molluscs and other aquatic animals, as might be inferred from the large and curved jaws. It is one of the air-breathing larvæ, and is therefore obliged to come frequently to the surface in order to take in a supply of air, which is done by means of a filamentous appendage at the end of the tail. Somewhere about July the larva has completed feeding; and then leaves the water and crawls up the bank, searching for a soft spot in which to burrow. Here it sinks its tunnel, and forms an oval cocoon, in which it awaits its change into the pupal state.

The Beetles belonging to this group, being, like the larvæ, air-breathers, are forced to come to the surface for the purpose of respiration ; and they contrive to carry down a supply of atmospheric air by enclosing a bubble under the bodies, where it looks like a globe of quicksilver as they swim about. As this species is not only handsome but harmless, it is in great favour with the keepers of aquariums, and is in consequence quite scarce in many places where it used to be plentiful, the professional dealers having ransacked all the streams within easy reach of London.

THE family next in order, the Sphæridiidæ, or GLOBE BEETLES, are distinguished from the preceding family by the shape of the tarsi, which are not fitted for swimming, and the hinder pairs of which members have the first joint much longer than the others. They are all small insects, rather globular in form, from which peculiarity the name of the family is derived ; and they are dark-coloured, black being the usual hue, relieved in some species by reddish spots. In the genus Cercyon, from which our example is taken, the club of the antennæ is large and bold, the palpi are slender, and the mentum is broad and flat.

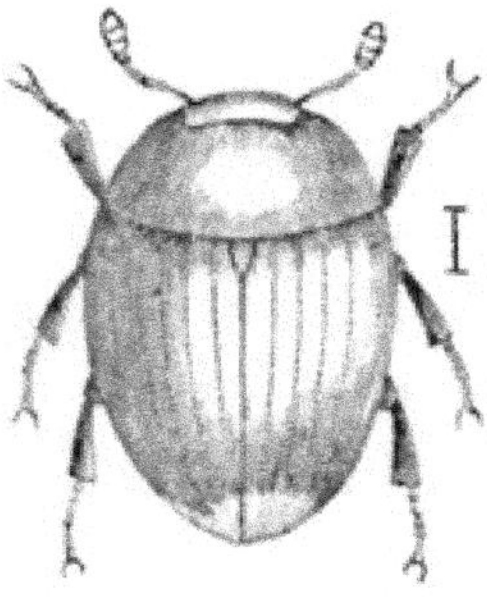

Cercyon anale.

As is the case with most red and black Beetles, the colour is exceedingly variable in different individuals, so different indeed, that the present species, *Cercyon anale,* a figure of which is herewith given, has been described under four separate names, each name being used to represent a different species. Its colour is black, but the elytra are generally tinted towards the apex with reddish-chestnut, the size and exact tint of this mark varying exceedingly; and in some specimens the elytra are altogether black. Like the rest of the genus, it can be found in cowdung.

Formerly some sixty species of Cercyon were catalogued, but they have now been reduced to seventeen genuine species; two-thirds of the imagined species proving, on careful investigation, to be nothing but varieties. In one case, that of *Cercyon nigriceps*, the same insect had been described under seven different names by the same naturalist.

CHAPTER VII.

LAMELLICORNES.

THIS chapter will be given to that very important group of insects, the LAMELLICORN Beetles, popularly called CHAFERS. The former term is composed of two Latin words signifying 'leaf-horned,' and is applied to these Beetles because the club of the antennæ is composed of a series of flat plates or leaves, which are movable like the rays of a fan, except in the Stag Beetle and its kin. The antennæ are always short, with a long or large basal joint, and set near the eyes and in front of them. If the reader should have skill to open an insect, he is strongly recommended to do so, in order to see the singular manner in which the large and apparently heavy bodies of these insects are lightened by a great number of air-vessels connected with the breathing tubes. These air-vessels extend all over the body, and are found even in the head.

The larvæ are fat, fleshy, soft-skinned grubs, feeding on vegetable matter, mostly, though not always, in a state of decay; and the last segment of the body is much larger than the others. After they are full-fed, they make cocoons from the chips of wood or other fragments of the material on which they have been feeding; and therein await their change into the pupal and perfect forms.

THE first family of the Lamellicornes is the Cetoniidæ, or Rose Beetle family. We have but few examples of these beautiful insects in England, and one or two of them are very rare. In this family the antennæ are short, and have only ten joints, three of them forming the club. The body is broad, and the elytra are flattened and not quite long enough to reach the end of the abdomen. A very familiar example of this family is the common ROSE BEETLE (*Cetonia aurata*).

Cetonia aurata.

This is a truly handsome insect. The upper surface of the body is beautiful shining-green, glossed with gold. The elytra have a number of impressed dots and curved marks scattered irregularly over them, and towards the apex are a number of scattered whitish marks, very variable in size, hue, number, and shape, according to the individual insect. Below, it is bright polished-copper.

The perfect Beetles are generally to be found in roses, especially the white and wild roses, which they are thought to damage in some degree. The larva, however, does not content itself with such light diet as rose-leaves, but lives on the less poetical but more substantial nourishment afforded by decaying wood, in which it remains for three years.

Sometimes, on opening a wood ant's nest, one of these larvæ is found in it, a circumstance which has gained for it the name of King of the Ants. It most probably takes advantage of the large quantity of wood-chips, bits of straw, fir-leaves, and similar material, with which the ants make their nest ; and finds therein an abundant supply of food. The ants do not meddle with it, probably finding that it remains in one spot, and does not interfere with the conduct of their nest.

After it has remained in the larval state for the full period, it makes for itself a cocoon from the wood-chips or other materials on which it has been feeding ; and it sometimes happens that when a decayed tree-trunk is suddenly blown or cut down, a great quantity of these cocoons roll out from among the ruins. Even when the larva has taken up its abode in an ants' nest, it makes a cocoon from the surrounding materials.

NEXT we come to the small though important family of the Melolonthidæ, which includes those insects which are popularly called Cockchafers. Only five species of this family inhabit England, and of these only two are even tolerably common. These two, however, more than compensate by their enormous numbers for the paucity of the other species ; and in some seasons are so exceedingly plentiful that they become an absolute pest to the agriculturist, laying waste thousands of trees, and destroying acre upon acre of pasture land.

Only too familiar to many of us is the common COCKCHAFER (*Melolontha vulgaris*). The insect is so well known that a detailed description is scarcely necessary. The peculiar bent projection at the end of the abdomen is worthy of notice, as are the rows of triangular white spots along its sides. There is a greyish down on the breast, and the elytra are covered with a yellowish down. Unless the insect have quite newly emerged from the pupal state and been handled very carefully, the down is sure to be rubbed off, and the beauty of the specimen greatly impaired ; so that a really perfect specimen even of so common a Beetle is worth preserving.

The life history of this insect demands a brief notice. The female deposits her eggs in the ground, where in due time they are hatched, and straightway begin to feed upon the roots of grass, which form the chief part of their diet. They remain in the ground for three years, continually destroying grass roots, and increasing to a wonderful size ; so large and fat, indeed, that their tightened skin seems scarcely able to hold its contents. The quantity of roots consumed by one of these insects is very great ; and in some places they have so completely destroyed the grass, that the turf

Melolontha vulgaris, larva (three parts grown).

has been completely detached from the ground, and might be rolled up by hand as easily as if the turf-cutter's spade had passed under it. These mischievous grubs do not confine themselves to grass-roots, but eat many of the underground crops, the potato often suffering terribly from them.

When full-fed, the grub makes for itself a cocoon in the earth, and then emerges, only to work as much destruction above the soil as it did below. In the larval state it fed upon the roots of grasses, and was out of sight; it now feeds on the leaves of trees, and is out of reach. In this way the Beetles are scarcely less mischievous than they were in their former state, for they will sometimes denude whole tracts of trees, so that, in the full beauty of summer-tide, the trees look as if the season were the depth of winter. In this country we are almost ignorant of the harm which the Cockchafer can do, for, although our crops and potatoes often suffer severely from its attacks, they are not wholly ruined, as is the case on the Continent.

THE family of the Geotrupidæ has eleven joints in the antenna, of which three form the club, and the margin of the head divides the eyes somewhat like the structure of the Gyrini, except that, in the case of those insects, the eyes are divided by a broad, flat band, and in the present family by a narrow ridge of horny substance. The body is very convex and the thorax large, in order to give room for the muscles that move the large wings and the powerful digging fore-legs.

One of the commonest English species, *Geotrupes stercorarius*, popularly known as the CLOCK, the DOR BEETLE, the FLYING WATCHMAN, the DUMBLE-DOR, and similar names, according to the locality in which it lives.

The colour of this species is black above, sometimes glossed with green or blue, and rich shining-violet beneath. On the middle of the clypeus there is a sharp tubercle. The thorax is smooth, except at the margins, which are thickly punctured, and on each side there is a nearly circular impression, thickly punctured in the interior. The middle of the scutellum is punctured, and the elytra are striated, the spaces between the striæ being smooth. The sexes may be distinguished by means of the tibiæ of the first pair of legs and the femora of the hind pair, the male having on the inner side of the front tibia a single erect spine, and the inner edge of the hind femora strongly toothed. As this insect is liable to much variation in colour, it is necessary to call attention to these minute points of structure by which the species can be definitely ascertained.

The life history of this Beetle may be briefly told as follows :—

In the autumn evenings the Beetles may be seen flying about in large circles, as if they were predacious insects quartering the ground in search of prey. In one sense, this is exactly what they are doing, as they are hunting after a favourable spot wherein to place their eggs, and are wheeling over the ground in hopes to find one. Attracted probably by the scent, the Beetle discovers a patch of cowdung, alights near it, crawls upon it, and straightway burrows through the soft material, and is lost to sight. When she—for it is the female who does the work—reaches the earth, she does not cease to burrow, but goes on with her labour until she has excavated a perpendicular tunnel some twelve inches in depth, and carried a quantity of the cowdung into it. In this substance she deposits an egg, crawls out of the burrow, and proceeds to make another, and so goes on until she has laid all her eggs.

The egg remains in its concealment until it is hatched, and then the larva consumes the food which its mother has taken the trouble to bring down for it. After this is eaten, the grub is strong enough to ascend the burrow and obtain as much food as it wants at the entrance. Within this retreat the larva passes through its transformations, and then ascends to the outer air, ready to take its part in the work of preparing nurseries for a future progeny. Seven species of Geotrupes are known in England. Twice as many species have been described, but recent investigations have shown that exactly half the supposed species were simple varieties.

In the accompanying woodcut is represented a Beetle of a very odd appearance, the sides of the thorax being prolonged into a pair of very formidable horns, a shorter horn occupying the centre of the anterior margin. This is the male *Geotrupes typhœus*, a near relative of the preceding insect. The female has only the veriest rudiments of horns, the anterior angles of the thorax being merely developed into a short, sharp prominence, like the teeth of a saw, while the place of the central horn is taken by a ridge running across the forehead. Indeed, owing to the absence of these horns, the female is so unlike the other sex, that no one who was ignorant of entomology would be likely to believe the two creatures to be nothing more than different sexes of the same insect.

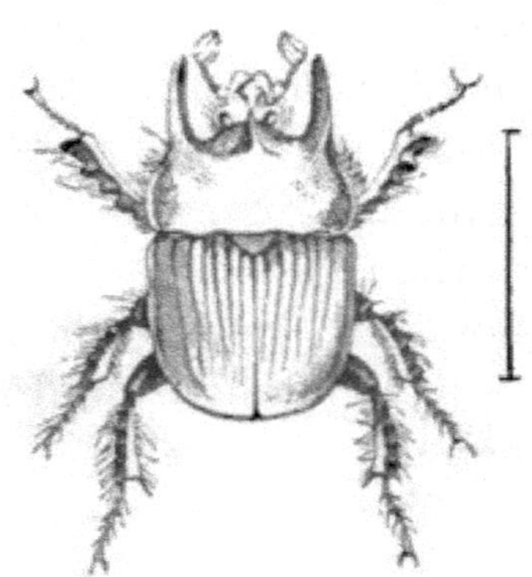

Typhœus fumatus.

The name of Typhœus was given to this insect in the days when classical mythology was the source of new names for insects, and to this species the name of Typhœus was given on account of its menacing aspect, which was fancifully compared to the giant Typhœus, who frightened Zeus and the other gods out of heaven merely by his looks. As, however, was the case with its ancient prototype, the Typhœus is not nearly so terrible as it seems, and its array of horns cannot do the least injury to the hands of its capturer.

There is no possibility of mistaking this insect, which, in addition to the characteristics that have been already described, has the first and last joints of the tarsus of equal length, and longer than the others. The colour of the insect is shining-black, the elytra being regularly but not very deeply striated. There is, however, considerable variation in this insect, as even in some males the horns are comparatively small, and little better developed than those of the female.

Considering that the Typhœus is a very common insect, it is surprisingly little known, and I have often found that entomologists who restricted themselves to the Lepidoptera alone have been totally ignorant of its existence, and expressed much surprise when I showed them a fine male Typhœus. This insect is in one sense an extremely interesting one, inasmuch as it is one of our few British examples of the strange and almost grotesque forms assumed by male Lamellicornes, but which are seldom seen except in exotic Beetles.

This insect may be found in all parts of England, being generally plentiful enough. The end of May and beginning of August are the times when it is in best condition, and at the latter season it may often be seen lying dead in roads or pathways. Like the Dor Beetle, it is a burrower, and has been taken while making its tunnel in sand.

THERE is a large family of small Beetles which must not be passed over without a brief notice. This is known by the name of Aphodiidæ, a name formed from two Greek words which refer to the habits of the different species, which are always to be found in cowdung. They are rather oblong and cylindrical in shape, and the elytra cover nearly the whole of the body. Many species are among the commonest of our British insects, and they must be sought in their accustomed haunts if the beetle-hunter wishes to obtain a good series of specimens. A stick will generally suffice to eject them from their hiding-places, but the 'digger' which has been already described is a more effective instrument, as the insects can be dug out of their shelter beneath the surface of the earth. Forty species of Aphodius are known to entomologists.

THE family of the Lucanidæ is represented in England by three genera, but by very few species, only one British species belonging to each genus. This family may at once be known by the club of the antennæ, which, though formed of a series of plates, is unlike that of the other Lamellicornes, in that the plates cannot be folded together, and are arranged so as to look as if the club were simply pectinated.

In the males the mandibles are enormously developed, and, in consequence of these distinctions, many entomologists have thought that the Lucanidæ ought not to belong to the Lamellicornes, but to be formed into a group by themselves. The name of *Pectinicornes*, or 'combed antennæ,' has been suggested as an appropriate title. As to this proposed alteration, Mr. Rye very judiciously remarks, that such a change ought not to be made until all the known Lamellicornes have been dissected, and their nervous system examined.

The genus Lucanus is distinguished by the flattened body, the apparently pectinated four-jointed club of the antennæ, and the enormous jaws of the male, which are often half as long as the head, thorax, and body together. Our only British species is the well-known STAG BEETLE (*Lucanus cervus*), which derives its popular name from the jaws of the male, which look somewhat like the horns of a stag.

In some parts of the country it goes by the name of HORN-BUG.

This is the largest of the British Beetles, as it sometimes attains a length of nearly three inches. The size, however, is extremely variable, as some males are barely half that length, and have their jaws comparatively small and weak. These are generally called undeveloped males, their inferiority being probably due to a want of food while in the larval state. Lest, however, a small and degenerate race of Stag Beetles should be perpetuated, the males always fight for possession of the females, and the consequence is, that none but the largest and strongest individuals have a chance of obtaining a mate.

The head and thorax of the Stag Beetle are black, profusely punctured. The elytra are deep-chestnut, becoming black on the margins, and at first sight appear to be quite smooth, but are, in reality, covered with the finest imaginable punctures. The jaws are of the same colour as the elytra, and the legs are black. The female is shaped like the male, with the exception of the jaws, which are small, curved, and sharply pointed. The head, too, is smaller than that of the male, because the muscles attached to the jaws are comparatively small.

This Beetle is in some parts of England very common, and in others not only rare, but absolutely wanting. I hunted insects industriously at Oxford for a series of years, and not only never saw a living Stag Beetle within many miles of that city, and believe that a specimen had never been taken in that locality. There is no apparent reason why it should find that

Oxford does not suit it, for the same trees flourish there as they do in Kent, where it is one of the commonest of the Beetle tribe, and the same water that flows past Oxford rolls through the Thames valley of Kent. Whatever may be the reason, the fact exists; and I well remember my gratification and astonishment when I first saw the Stag Beetles flying about nearly as plentifully as Cockchafers or Dor Beetles.

The larva of this insect somewhat resembles that of the Rose Beetle, and lives in rotten wood.

The oak supplies its favourite food, but it also lives in the willow; and, according to some entomologists, the willow-fed specimens are smaller than those which live in the oak. These larvæ often do very great harm, their powerful jaws enabling them to eat into the living as well as the dead wood, and into the roots themselves. It remains in the larval state for at least four, and perhaps as much as six years, and when it is about to become a pupa, makes for itself a cocoon out of the wood-chips with which it is surrounded.

The jaws of the male are quite as formidable weapons as they appear to be, the muscles which close them being very powerful, and their sharp and strong teeth inflicting a severe bite. Mr. Curtis mentions that the jaws retain the power of biting long after the head has been separated from the body, and that in one case when a severed head of a Stag Beetle was taken home in the evening, it retained on the following morning sufficient power to pinch the finger. Still, severe as is the bite of the male

Stag Beetle, that of the other sex is still more severe, the points of the strong, sharp, curved jaws being made to meet in the flesh.

At first sight it would appear that the insect must be a carnivorous one, and that such formidable weapons were used for the purpose of capturing and destroying other insects. In reality the Stag Beetle is essentially a feeder on juices, which it obtains by wounding twigs and fruits with the sharp teeth of its mandibles. If kept in captivity, it will feed on moistened sugar, and has a curious way of flattening itself on the ground, in order to reach the sugar with its tongue. Indeed, it only uses its jaws as weapons of offence, when it fights for the possession of the female, or when it is captured and wishes to escape. It will bite fiercely in such a case, and, if kept alive, will resent with open jaws any attempt to disturb it.

CHAPTER VIII.

STERNOXI.

THE group that now comes before us is a very boldly marked one, and is known by the title of *Sternoxi*, or 'sharp-breasted,' because the prosternum, or under side of the thorax, is prolonged backwards with a sort of spike, which fits into a cavity between the middle pair of legs. This projection is technically named the 'mucro,' or dagger. The body is long, rather cylindrical, but slightly flattened, and the antennæ are mostly serrated, but sometimes pectinated, and in a few instances nearly plain and thread-like. There are other distinctions, but these are so bold and evident that they will be quite sufficient for the reader's purpose.

Most of these insects possess the curious power of leaping, which has earned for them the popular title of SKIPJACK BEETLES. Their legs are very short, so that if the Beetle should by any chance fall on its back on a flat surface, it would have no power of recovering itself, but for the curious piece of mechanism of which the 'mucro' forms a portion. Whenever the Beetle falls on its back, and cannot recover itself, it lies still for a few moments, and then begins to arch its body, so that it rests only upon the end of the abdomen and the back of the head, the thorax being well elevated. By this movement, the mucro is drawn out of the groove into which it fits. Suddenly, the insect reverses its position and springs the elastic mucro into its place, thus driving the base of the elytra against the ground, and causing itself to fly up into the air.

The spring is always accompanied with a slight but sharp clicking sound, from which these insects have derived the name of CLICK BEETLES. There is an absolute necessity for this curious provision of nature. The Click Beetles are all feeble, slow, and defenceless, and their only way of escaping from an enemy is by loosening their hold of the herbage on which they are crawling, and allowing themselves to drop to the ground. The sweep-net is very useful in catching these Beetles, as it anticipates the movement, and captures them as they fall.

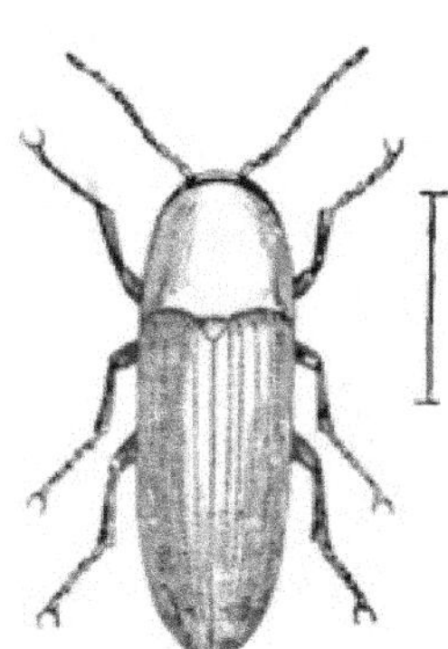

Elater san guineus.

WE will now proceed to describe one or two typical examples of this group.

Of the typical genus our example is *Elater sanguineus*, a figure of which is here given. In this family the antennæ are long, and inserted just in front of the eyes, which are large and round. The two hinder angles of the thorax are produced into spines pointing backwards, and the mucro is able to move freely in the cavity into which it fits, in consequence of the distance between the base of the elytra and the thorax. In this genus the tarsi are bristly, the joints becoming gradually smaller, the body is flattened, and the sides of the thorax are not widened. The name *Elater* is Greek, and signifies a 'striker' or 'hurler,' the name being given to the insect in consequence of its power of hurling itself into the air.

The ground colour of the present species is black, but it is covered with brown-black or red-brown down, the thorax being rather convex, punctured, and having a short and shallow furrow behind. The elytra are of a more decided hue, being blood-red (whence the specific name, *sanguineus*), and are striated and punctured. It is not a very common insect, but may be found in woods. I have taken it in a copse on the Wiltshire downs.

The larvæ of the Skipjack Beetles are popularly known as Wire-worms, a title which has been applied to them for two reasons—first, because they are long and narrow, seeming to be drawn out, as it were, like wire; and, secondly, because they are tough and hard-skinned, so that a roller passing over them does them no manner of harm, but only squeezes them into the soil, as if they were so many pieces of wire.

Some of these larvæ commit terrible ravages among the crops, feeding upon the roots, and so remaining themselves hidden while their ravenous jaws are destroying the very life of the plants. There are few things which irritate an agriculturist more than such a foe as this. He makes up his mind to the caterpillar, the turnip-fly, the snail, and other creatures which devour the plant itself. They at least can be seen while eating, however closely they may conceal themselves at other times, and the amount of food which they take is proportionate to the mischief which they do. But the Wire-worm wastes and eats in concealment, and, while it only eats one-tenth the amount of that consumed by a caterpillar of equal size, destroys ten times the number of plants. Various projects have been set on foot for extirpating the Wire-worm, but I hear of no plan that has succeeded except that provided by Nature, namely, the fondness of certain birds for the Wire-worm. Mr. Westwood mentions that even pheasants are useful to the farmer in this respect, their crops having been found stuffed with Wire-worms. There are very few plants or flowers which this voracious insect will not attack, and the gardener as well as the farmer is therefore interested in the Skipjack Beetles and their progeny.

The mole is a great eater of Wire-worms, as it finds them near the surface, and can take them while making the superficial burrows which it often excavates within an inch or two of the surface of the ground. It is stated that this destructive larva remains five years in the ground before assuming the pupal stage, so that we ought to encourage as far as possible every creature which assists in keeping down its numbers.

Our last example of these insects is that which is represented on the next page, and is known by the name of *Campylus linearis.*

As the reader may see by reference to the figure, this insect is very unlike any of the Beetles of this group which have been already described. The head projects boldly from the thorax, and the eyes are very large. The hinder angles of the thorax are rather elevated, sharp, and bent outwardly. The body is long and slender, a fact which has gained for the insect the specific title of *linearis*. The generic title *Campylus* is of Greek origin, the word signifying a peculiar staff; and the name has been chosen on account of the slender, stick-like form of the insect.

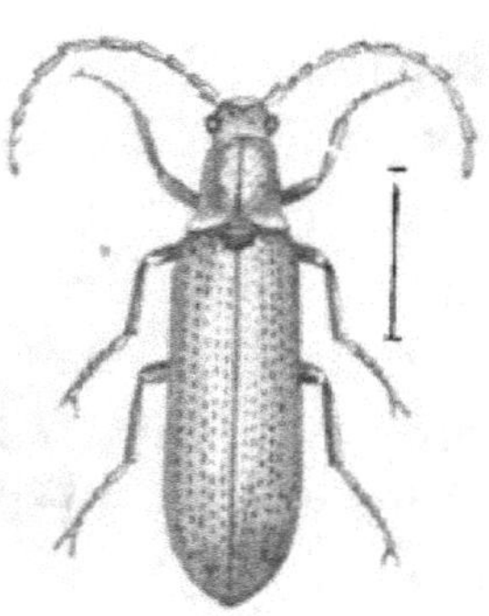

Campylus linearis.

This is an extremely variable species in point of colour, the female being very remarkable in this respect. Red, however, is the leading colour, and the average hues of the insect may be described as follows:—The head is black and deeply punctured, and the thorax has a deep furrow along the centre, and a transverse pit or 'impression' a little behind the middle. Its colour is brick-red, and in many specimens there is a black spot on the centre, while in others the same part is brown. The elytra are rather more convex in the female than in the opposite sex, and are covered with striæ and punctures. Their colour is somewhat the same as that of the thorax, but rather paler, though in many examples, especially among the females, the whole elytra are yellowish brown, except the margin, which retains the ordinary brick-red colour.

The insect is a plentiful one, especially in certain years. It is best taken by means of the sweep-net, which should be used along the sides of hedges, in copses, and similar localities. It is the only British example of its genus.

CHAPTER IX.

MALACODERMI.

IN this group of Beetles are gathered together a number of Beetles differing much from each other in many points, but agreeing in the one characteristic which has gained for them the name of MALACODERMI, or Soft-skinned Beetles. In these insects the exterior of the body, instead of being quite hard and strong, as is the case with those Beetles which we have examined, is soft and flexible, and generally covered with a very short and delicate down.

We begin our notice of these Beetles with the family of the Lampyridæ, of which only one species inhabits England, namely, the well-known GLOW-WORM (*Lampyris noctiluca*). In this family, the female possesses neither wings nor elytra, the head is concealed under the large and rounded prothorax, and both sexes have the power of emitting a phosphorescent light, the lamp of the female being very much brighter than that of her mate.

This, almost our sole representative of the exotic light-giving insects, is fortunately very plentiful in this country, and may be seen abundantly in sheltered spots, preferring those which are slightly damp. It is very abundant in Kent, and in the summer evenings the green-blue lamp of the Glow-worm may be seen shining amid the leaves. If examined in the dark the light is seen to proceed from the three last segments of the body, the under side of which emits the light in a wavering, uncertain sort of way, the fact of being handled seeming to alarm the insect and cause it to retain the light-giving power. Sometimes, indeed, it puts out its lamp altogether when handled, the light being evidently under the control of the insect. It is said, however, that if a Glow-worm be placed in oxygen gas the light is greatly intensified, and the Beetle seems unable or unwilling to retain it. Gilbert White, in his 'Selborne,' remarks that the Glow-worms put out their lamps between eleven and twelve P.M., and shine no more for the rest of the night.

The dissimilarity between the sexes of the Glow-worm is very strongly marked, the female being entirely wingless, while the male has large wings and elytra which cover the whole of the body. It is popularly thought that the male does not possess the light-giving power; but this is a mistake, as every practical entomologist must know. Still, though the male does possess a lamp, it is very much smaller and feebler than that of the female, and, instead of a mass of phosphorescence, throwing a radiance of some inches in extent, it is nothing more than two tiny spots of light, no larger than minnikin pin's heads. I once

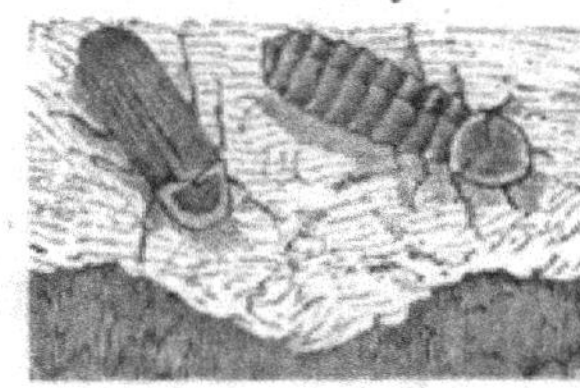

Lampyris noctilus (male and female).

took a male Glow-worm on the wing with his lamps lighted.

As to the object of the light, it is a matter of very great uncertainty. The obvious solution of the problem is to say that the light is intended to guide the male to his mate.

I cannot of course say whether this is the case or not, but I do not see that this is the sole object of the light. There are plenty of night-flying insects which manage to find their mates in the dark without the use of any such aid, being attracted to them by scent rather than sight; and, even if the light emitted by the female Glow-worm be intended for such a purpose, that of the male cannot be of the slightest use either to him or to the mate whom he is seeking.

Moreover, not only the perfect insects, but the pupæ, the larvæ, and even the eggs are slightly luminous, so that in these cases the light evidently cannot act as a guide. I am inclined to believe that no utilitarian theory will account for this singular development of light from a living insect, and that the phosphorescence was given to it for the same reason that the butterfly's wing glows with many-coloured plumage, and the rose is dowered with softly-tinted petals and sweet perfume.

This insect is doubly interesting to the entomologist. In the first place, it is a British light-producer: and in the second, its life in the larval state is a very valuable one to the agriculturist. It feeds on snails, attacking and devouring them while still alive, their shells being no protection to the luckless molluscs. The structure of this larva is rather remarkable. In the first place, it bears a singularly close resemblance to the perfect female insect; and in the next, it is furnished with a peculiar apparatus at the end of the tail, which serves a double purpose, primarily of assisting in locomotion, and secondarily acting as a brush, by which the slime of the snail can be swept from its body. In some works on entomology, this organ has been erroneously drawn like a shaving-brush cut off square at the end, whereas it consists of some seven or eight projections from the end of the body, which can be protruded or withdrawn at will. Almost as soon as the snails begin to come out from the hiding-places in which they have lain dormant through the winter, the Glow-worm larva is ready to attack them, and thus plays its part in reducing the number of snails that would have been produced by those which it kills, and so helps to preserve the balance of Nature.

The generic name of *Lampyris* is formed from two Greek words signifying 'shining-tail.'

THE family of the Telephoridæ comes next in order. These insects have long and very soft elytra, which often do not cover the whole of the abdomen. The head is not hidden under the thorax, and both the antennæ and the palpi are slender. The various species are very plentiful, especially on the flowers of umbelliferous plants, and are popularly known as Soldiers and Sailors—the red species being called by the former name, and the blue species by the latter.

One of these Beetles, called *Telephorus fuscus*, is shown in the accompanying woodcut. In this genus the elytra reach to the end of the abdomen, and the thorax is not notched. Soft-bodied as are these Beetles, they are among the most quarrelsome of insects, and fight to the death on the least provocation. Indeed, it has long been the custom for boys to catch these Beetles, and set them fighting with each other. There is not the least difficulty in this, inasmuch as the Beetle is as ready for battle as a game-cock, and, not content with fighting to the death, eats its vanquished antagonist after killing it. The popular idea among boys used to be, that a soldier and a sailor must be pitted against each other; but this is not the case, for these Beetles will fight and devour each other without the least reference to species or even to sex, so that a soldier male and female will fight as fiercely as if they were two males of different species.

They are not active insects, and though they can fly well, and use their wings freely, are slow of progress, and can be taken by hand while in the air. Like the perfect insect, the larva is carnivorous, feeding generally upon earth-worms, but having no scruple in devouring its own kind. These larvæ may be found among grass and moss during the earlier months of the year, after the severe frosts have ceased. They pass the whole of the winter in the larval state, and assume the pupal condition about April or May, according to the warmth of the season. Twenty-six species of Telephorus are indigenous to England.

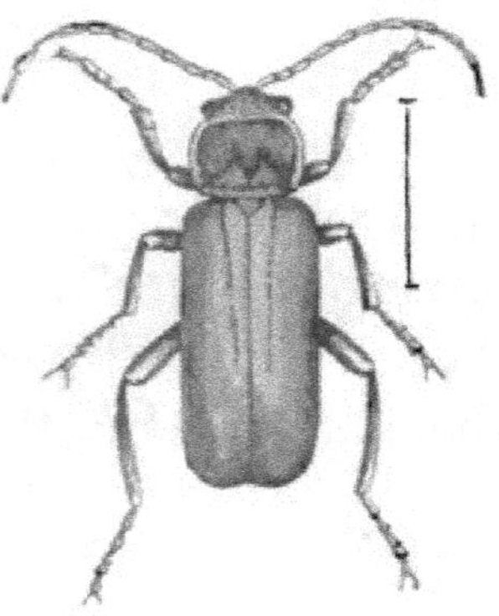
Telephorus fuscus.

THE family of the Cleridæ are mostly beautiful insects, and although they are not large, some of them rank among our prettiest Beetles. The body is oblong, hard, and covered with down, and the head and thorax are not as wide as the elytra. The genus Clerus is known by several points of structure. The basal joint of the tarsus is very minute, the last joint of the labial palpi is hatchet-shaped, and the last joint of the antennæ is large, rounded, and furnished with a curious projecting point directed inwards.

In their larval state these Beetles are carnivorous and parasitic on other insects. We can take one example of this pretty genus only, named *Clerus formicarius*, which is shown in the accompanying woodcut. The head of this insect is black, and the thorax brick-red, the front margin being black. The elytra are very boldly coloured, their ground hue being black crossed by two snow-white bands, shaped as seen in the illustration, and their base is of the same colour as the thorax. The larva is found under the bark of trees, not to eat the wood or bark, but to destroy and feed upon the larvæ of wood-boring Beetles. Its colour is dark-pink, spotted in front. The specific name of *formicarius* is given to this Beetle because it has an ant-like aspect.

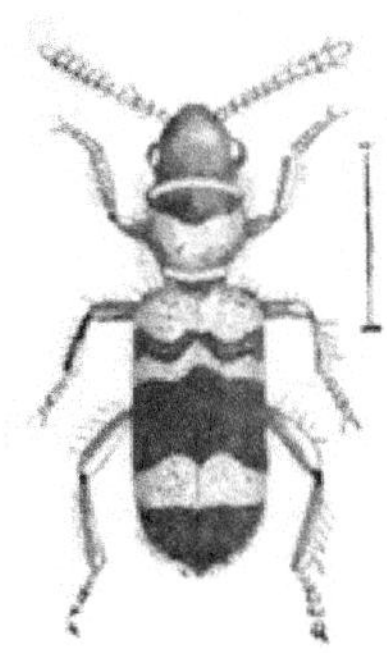
Clerus formicarius.

THE reader will remember that the larva of the first example of the Cleridæ feeds on those of certain wood-boring Beetles. We now come to one of the insects which furnishes unwilling nourishment to the Clerus, which is called *Anobium striatum.* This Beetle belongs to the family of the Ptinidæ, a group of small and very destructive Beetles. They are cylindrical in shape, covered with very short down, and are able to draw their heads completely under the overhanging thorax. Their legs can be packed closely to the body, and the tarsi have five joints. The genus Anobium, of which there are eleven British species, has the three last joints of the antennæ rather longer than the others, and the last joint egg-shaped.

The various species of this genus work terrible havoc among furniture, in which they produce the defect that is popularly known by the name of 'worm-eaten.' They are not in the least particular as to their diet, and will devour almost any kind of food.

For these troublesome little pests I know of but one remedy, namely, injecting into the holes spirits of wine in which corrosive sublimate has been dissolved. This is not so tedious a business as it may seem to be, as the spirit will often find its way from one hole to another, so that, if half a dozen holes be judiciously selected, the poison will penetrate the whole piece of wood, kill all the insect inhabitants, and render it for ever impervious to their attack. A New Guinea bow in my collection was riddled with the burrows of the Anobium, but was easily cleared of its inmates. Holding the bow perpendicularly, I injected the spirit into several holes at the upper end. The effect was magical. The little Beetles came out of the holes in all directions, and not one survived the touch of the poisoned spirit; many of them, indeed, dying before they could force themselves completely out of the holes. They will also eat skins and any dried animal substance; and I have found a neglected box of moths completely destroyed by these voracious insects.

The present species is rather convex, and blackish-brown in colour. The thorax is rather narrowed behind, and on each side of the hinder margin are two pits. The elytra are boldly striated, each stria being seen, when examined with the aid of a lens, to consist of a number of punctures placed in regular rows. It is a very common species.

The old popular terror respecting the Death-watch is well known, a mysterious ticking being heard in the dead of night, which was—and is still—supposed to presage the approaching death of some one in the house. The ticking of the Death-watch is, in fact, the call of the Anobium to its mate, and, as the insect is always found in old wood, it is very evident why the Death-watch is always heard in old houses.

OUR last example of this group is here represented, the sketch showing the profile of the insect, whose name is *Mezium sulcatum.* There are three insects very closely resembling each other, belonging respectively to the genera Mezium, Gibbium, and Niptus, each being the sole British representative of its genus. The two former are almost exactly alike, but can be distinguished by looking at the thorax with a lens, the difference being that in Gibbium the

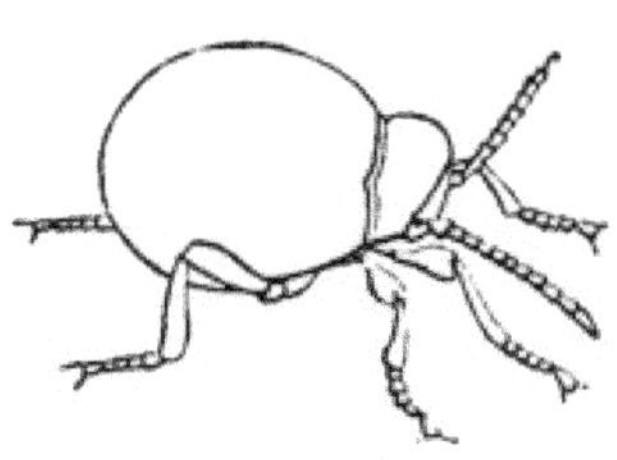

Mezium, side view.

thorax is smooth, whereas in Mezium it is covered with longitudinal furrows, whence the name *sulcatum*, or 'furrowed.'

To my mind, these are the oddest-looking Beetles that we have in England, and, indeed, at first sight they much more resemble spiders than Beetles.

The Mezium can generally be found in the cupboards and other recesses of old houses, and, indeed, all three Beetles may be captured on the same premises. A very good trap for them is a deep and steep-sided basin, with a little moist sugar at the bottom, and a stick or two laid against the sides by way of a ladder. The Mezium is very fond of sugar, climbs up the ladder, lets itself tumble into the sugar, and then cannot get out again, not being able to cling to the polished sides of the basin. It is thought by many entomologists, that neither of these Beetles is indigenous, but that all three have been imported from abroad.

CHAPTER X.

HETEROMERA.

THIS is a very important section of the Coleoptera, embracing many of our most familiar Beetles, though in England the number of Heteromerous Beetles is very small when compared with the list of exotic insects. The name *Heteromera* is compounded of two Greek words, signifying 'unequal-jointed,' and is applied to these Beetles because they all have five joints in the tarsi of the first and intermediate pairs of legs, and only four joints in those of the hinder pair. As has been mentioned in connection with other four-jointed tarsi, the full number of five joints is in reality present, but the basal joint is very long, and in it is merged the missing joint; so that the joint is in reality not absent, but so small as to escape ordinary observation.

The first family of this section is named Blaptidæ, and its members are known by the 'securiform,' or hatchet-shaped last joint of the maxillary palpi, and the long femora of the hind legs. The wings are not developed, and the elytra are soldered together. In England we only have one genus of this family, containing three species. That which we will take as our type is the CHURCHYARD BEETLE (*Blaps mortisaga*), which is represented below. All the species belonging to this genus are—to use a word which I do not like to apply to insects—ugly. They are dull, dead-black in colour, are wonderfully sluggish, crawling slowly as if afflicted with rheumatism, and always frequenting damp, dark, and dismal places. Being often found in the murkiest crannies of cellars, they have gained the popular and appropriate title of CELLAR BEETLES.

The species represented in the illustration may be recognised by the bold puncturing and contracted base of the thorax, and the lengthened projection at the apex of the elytra. It is not so generally plentiful as the second species, *Blaps mucronata*, being seldom found in the southern parts of England. Still these species are very similar in their habits. They are possessed of a very nauseous odour, suggestive of dwelling among the graves. Yet, unpleasant as these Beetles may be, we are informed that an Egyptian species, *Blaps sulcata*, is employed as a remedy for ear-ache, and a cure for the sting of the scorpion; while the women are in the habit of seeking and eating it, in order to produce the fatness which is thought in the East to be an essential point in female beauty.

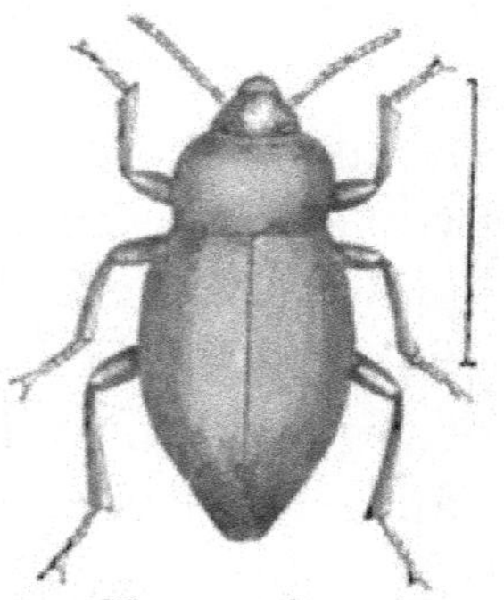

Blaps mortisaga.

THE family of the Tenebrionidæ only contains one genus, and that genus but two British species.

These Beetles possess large wings, and are capable of flight, the elytra not being soldered together as is the case with the Blaptidæ. The thorax is squared, and its base is as wide as the base of the elytra.

Herewith is represented the typical British species, *Tenebrio molitor*, the colour of which is shining blackish-brown, the body being rather flat and very thickly punctured. Each of the elytra has one very short stria next to the scutellum, and eight others reaching to the apex. It lives in corn-mills, flour-stores, bakehouses, and similar localities, and in consequence is often called the FLOUR BEETLE.

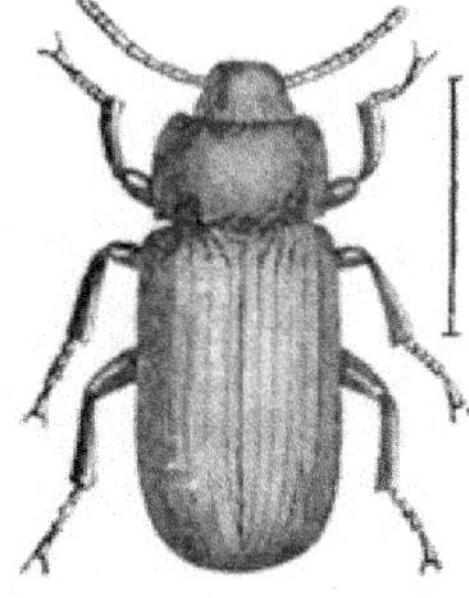
Tenebrio molitor.

The larva is popularly known by the name of MEAL-WORM, under which name it is largely supplied to bird-fanciers, who find that many of their feathered pets will not live unless they have a constant supply of insect food, such as is afforded by the Meal-worm, which, in consequence of the perpetual warmth of its home, breeds throughout the year. From April to June is, however, the best time to find the perfect insect.

THE family of the Pyrochroidæ is rendered familiar to us by means of the well-known CARDINAL BEETLE (*Pyrochroa rubens*), so called on account of its beautiful scarlet colour. The insect is represented below.

The Pyrochroidæ are known by the distinct neck the rounded thorax, and the form of the antennæ, which in the males are boldly toothed. The mandibles are deeply notched at the tips, the maxillary palpi have the last joint rather axe-shaped, and the elytra are long, wide, and cover the whole of the abdomen. The typical genus has the antennæ longer than the head and thorax, and very boldly 'pectinated,' or comb-like, in the males, in which sex the eyes are distant from each other. 'Pectination' is nothing more than a development of 'serration,' or saw-like form, each of the joints being drawn out into a long and narrow tooth, sometimes on one side only, but often on both sides. The latter form of pectination is conspicuously shown in many moths, as we shall see when we come to treat of these insects.

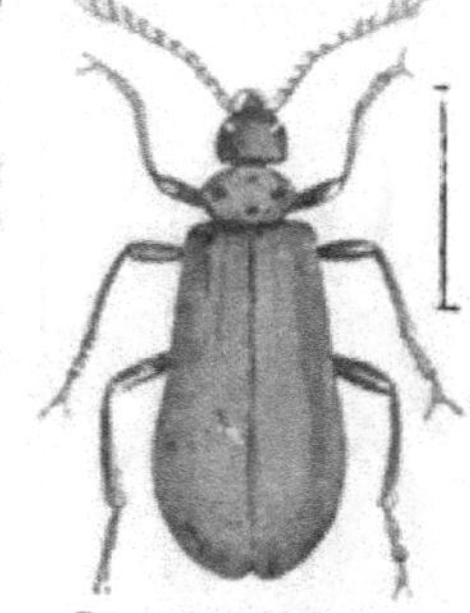

Pyrochroa rubens.

The forehead of the Cardinal Beetle is black, and there is a curved rust-red mark between the eyes. The thorax and elytra are rich scarlet, intensified by a short velvety down with which the surface is covered. This insect is as plentiful as it is handsome, and it may be captured throughout the summer. It is often one of the inmates of the sweeping-net, after that implement has been used among the flowers of hedgerows. The larva is whitish in colour, and inhabits decaying willows.

EVERYONE who has walked in the country, and used his eyes, must have noticed the well-known OIL BEETLES (*Meloë*), so called from their curious habit of ejecting a drop of clear yellowish oil from the joints of their legs when they are handled. The colour of these Beetles is dull, dark indigo blue, and they are wingless, slow-moving insects, especially the females, so that they have no chance of escaping from capture, to which their very conspicuous shape renders them liable.

Meloë cicatricosis (female).

The life history of the Oil Beetle is a very curious one. The female Beetle deposits in little holes in the ground a vast number of the tiniest imaginable yellow eggs, placing several thousands in each hole. As soon as the eggs are hatched, the larvæ make their way into the open air. They are most extraordinary creatures, and no one who saw the newly-hatched and the full-grown larva of this Beetle would ever imagine that they could be the same creature, and in the same stage of metamorphosis. They are scarcely so large as the semicolon (;) used in this work, and are long-bodied, furnished with six long and prehensile legs, and gifted with great activity. As soon as they reach the open air, they climb the stems of flowers and gain the blossoms, where they lie in wait. Presently a bee comes to gather honey or pollen, when the little larva leaves the flower, climbs upon the bee, and clings to its body with its clasping legs.

Meloë, young larva, magnified.

The bee, unconscious of its new burden, goes as usual to its nest, when the larva quits its hold, and remains in the nest. The parent bee being gone, thinking that everything is right, the Meloë larva devours the egg, and then throws off its first larval form in order to assume another, in which it somewhat resembles the grub of the cockchafer. It now turns its attention to the food prepared by the bee for its young, and finds therein just sufficient nutriment to carry it through its larval condition. The reader will see that it bears not the least resemblance to the long-bodied, quick-legged larva in the first stage of growth.

To prepare these insects for the cabinet requires some little care and patience, especially with the females, for when the creature dies, the large soft abdomen begins to shrink, and when it is quite dry, the abdomen is not one-third its proper size, is full of wrinkles, and crumpled out of all shape. The only plan, therefore, is to stuff it with cotton wool. The usual mode of so doing is, to cut a slit on the under side, remove the contents of the abdomen, and replace them with cotton wool. I have, however, found this plan scarcely satisfactory, inasmuch as the edges of the slit are apt to recede from each other, so that the cotton wool is visible. There is another plan, certainly involving more trouble, but with far better results. With sharp scissors cut off the abdomen altogether, squeeze and draw out its contents gently by the hole which is made at its base by the blades of the scissors. Through the same aperture introduce the cotton wool, a very little at a time, so that you can exactly restore the original shape of the abdomen, taking care to stuff it a trifle larger than it was originally, because the skin will contract a little on the cotton wool. Now, stick the point of a needle perpendicularly into the setting-board, and pass the eye into the abdomen through the same hole, so as to prevent it from losing shape by lying down. Set the other half of the Beetle independently, and, when both parts are quite dry, join them with a tiny drop of coaguline. If this be properly done, there will not be the slightest mark of any junction, and the specimen will always look as well as it did when living, and preserve its soft, rounded contour.

CHAPTER XI.

RHYNCHOPHORA, OR WEEVILS.

THESE terribly destructive insects do not attain any great dimensions in England, but they make up for their diminutive size by their enormous numbers. How many species are known to inhabit this country it is impossible to say, as new species—especially those of small size—are continually being added to our lists; but if we say that about five hundred British species are at present catalogued, we shall be very near their number.

The name *Rhynchophora* is formed from two Greek words, signifying 'snout-bearer,' and is given to these insects because the head is very much prolonged and narrowed, in some species looking like the long curved beak of the ibis or curlew. The mouth and its accompanying organs are always at the end of this beak, and in some species of Weevils the resemblance to the head and mouth of the Porcupine Ant-eater of Australia is really startling. The name of *Tetramera*, or 'four-jointed,' was formerly given to this group, because its members appear to have only four joints in the tarsi. Mr. Westwood, however, with his wonted acuteness, pointed out that there were really five joints, the missing joint being microscopically small, and hidden under the lobes of the third joint. The three basal joints of the tarsus are always furnished with a thick pad beneath, and may be seen by examining the feet of any of our common Weevils with a pocket-lens.

The antennæ are always set well in front on the 'rostrum,' or beak, and in most, though not in all, species are furnished with a very long basal joint, so that they are elbowed, or 'geniculated,' according to the scientific term.

According to the system which is at present in vogue, the Weevils are divided into two sections—namely, those in which the antennæ are not elbowed, and those in which they are. The former are called *Orthoceri*, or 'straight-horned,' and the latter *Gonatoceri*, or 'knee-horned.' We begin with the former, and take for our first example of these Beetles the RED-FOOTED WEEVIL (*Bruchus rufimanus*), a figure of which accompanies this description. This insect belongs to the family Bruchidæ, which have antennæ rather serrated, and becoming gradually thicker towards the apex. The elytra do not reach to the end of the abdomen, and the basal joint of the tarsus is long and curved. In the genus Bruchus the antennæ are rather delicate, and the elytra are oblong and squared.

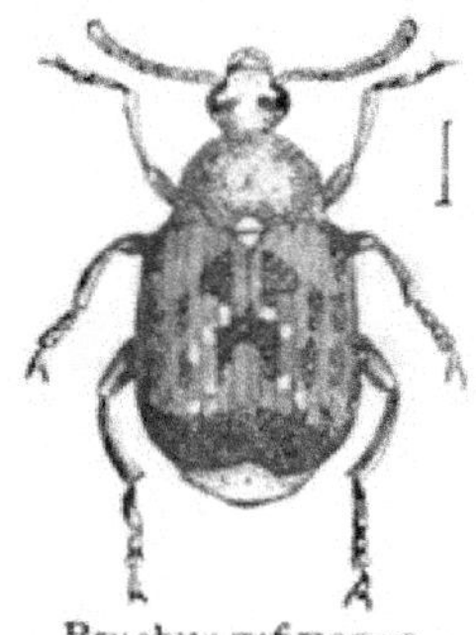

Bruchus rufimanus.

The Red-footed Weevil is rather variable both in size and colour, but is usually as follows:—The general hue is black. Upon the thorax, which has its edges slightly waved, there are two white spots upon the disc, and a large triangular spot about the middle of the base, some ashy-white hairs being scattered on the disc. The elytron is striated and punctured, and has a number of white spots scattered over it, and a whitish-grey streak near the scutellum. The tip of the abdomen is white, with the exception of two dun-coloured spots.

All the species of this genus are exceedingly destructive, feeding upon the seeds of beans, peas, and similar vegetables, very often doing enormous damage by dint of numbers, in spite of their small individual size. The reader may perhaps have had occasion to notice that, when peas are newly shelled, one frequently occurs in which a hole or a groove is scooped, the tenant being a little white maggot. These maggots are almost always the larvæ of this or some allied species of Weevil. They remain in the seeds until they have attained their perfect condition, when they escape through a round hole made for the purpose while in the larval state. Eight species are acknowledged to be indigenous to Great Britain. Many other species have been placed on the British lists, but entomologists have decided that they have been introduced into England in cargoes of peas, beans, or corn, and therefore ought not to be admitted as genuine British insects. Indeed, it could be wished that the law of extradition could be extended to insects, and that these Weevils, together with the cockroach and sundry other destructive and noxious insects, could be restored to the country whence they came.

THERE is an enormous genus of Weevils, consisting of very tiny species, about as large and somewhat the shape of a note of admiration (!) as here given. Their bodies look very much like pears, the stalk of the fruit representing the beak of the insect. In consequence of this resemblance they have received the generic name of *Apion*, which is a Greek word signifying a pear. They have also been compared to peg-tops; and, if the peg were curved instead of straight, the resemblance would be almost complete. We will call them Pear Weevils.

In spite of the difficulties which attend the examination of these insects, the entomologist will find his time well bestowed upon them. Independently of other sources of interest, these tiny Beetles are marvellously beautiful. Their colours are exceedingly various, and the richness and perfection of the sculpture which adorns their tiny bodies must be seen to be appreciated. It is as if the very exuberance of creative power had sported with these little creatures, a thousand of which could be contained in a lady's thimble, and yet which bear upon every portion of their bodies a limitless profusion of highly-elaborated ornament. The head and thorax are covered with a multitude of deep impressions, at first seeming as if scattered at random, but in reality disposed with most consummate art; while the elytra baffle all attempts to describe their varied beauty. Agreeing in one point—namely, the bold ridges which run longitudinally along them—they are of infinite variety in their details, so that a full description of all the species would occupy much more than the space that can be given to the whole of the Beetle tribe.

Tiny as they are, the Apions often do much damage to the agriculturist, many of them living in peas and beans, as has been stated of the Bruchus, some boring into the stems or roots of plants, or making a gall-like excrescence on the leaves or twigs. They specially frequent clover, and in a field of this plant, and along the adjoining hedgerows, the entomologist can take sufficient Apions in a morning to give him full employment during the winter months with his microscope. I may here mention that some knowledge of drawing is a potent help in the study of insects; and, indeed, the note-book and pencil should be always at hand. No matter how rude may be the sketch, it is sure to be useful, and has a wonderful power in fixing details in the mind.

In the accompanying woodcut is shown *Apion carduorum*, being about one-seventh of an inch long, while the generality of Apions are not much more than half that length. The head and thorax of this insect are black, with short shining hairs scattered very thinly over the surface. Near the base of the head the antennæ are set upon two rather bold tubercles. The elytra are of a verdigris-green, with a tinge of blue—a colour which is rather common to this genus—and the spaces between the striæ are very flat.

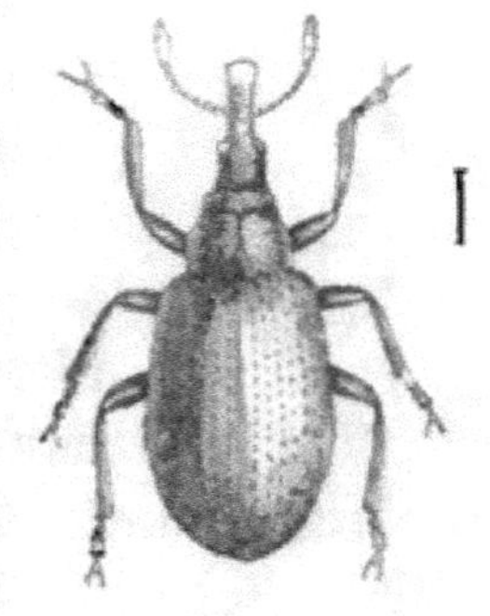

Apion carduorum.

Some eighty British species are known to entomologists, and I would strongly recommend the beginner to lay aside the examination of these little beings until his eye is trained to seizing details by some practice with the larger insects.

The specific name of this insect, *carduorum*, signifies 'of the thistles,' and is given because it can be found upon that plant.

WE now come to the Weevils with elbowed antennæ, the first family of which is the Brachyderidæ. In these insects the head is short, wide, and set on the thorax without any separate neck, a peculiarity which has gained for the family the name of *Brachyderidæ*, or 'short-necks.'

Our example of this family is *Sitones lineatus*, which is represented in the woodcut which accompanies this description. This genus is known by the possession of wings, the short beak, and the third joint of the antennæ, which is shorter than the second. About nineteen British species of this genus are acknowledged. The present species is a pretty though not a brilliant insect. The ground colour is black, but the body is clothed above with scales of a warm-brown hue, while the under surface of the body is similarly clothed, but with scales having a silvery lustre. There is a central furrow on the disc of the thorax, and a rather deep impression across its apex. The elytra are punctured and striated, with white interstices between the striæ. These white lines form the distinguishing characteristic

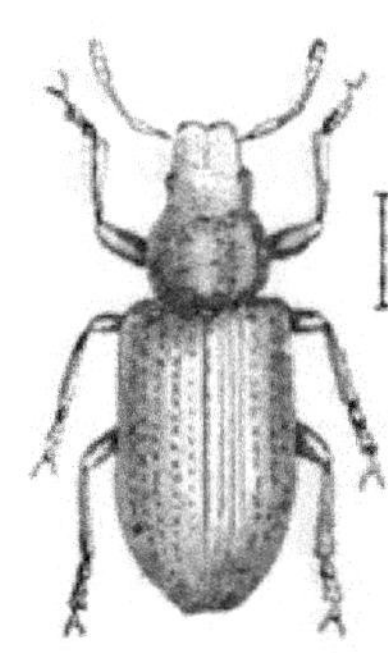

Sitones lineatus.

of the species, which is apt to be very variable in other respects.

The young entomologist must be very careful in handling this and other scale-clad Weevils, as the scales are easily rubbed off, so that nothing is seen but the dull-black of the elytra. A roughly-handled Weevil is just as unfit for the cabinet as a butterfly with the rich plumage rubbed from its wings.

All the members of this genus are injurious to the crops, especially clover and peas. The generic name of Sitones (erroneously spelt Sitona by some entomologists) is a Greek word signifying a corn-dealer, and has been given to the Beetles in consequence of their influence upon the harvests.

THE next family upon our list is that of the Otiorhynchidæ. This rather crabbed name is compounded of two Greek words, the former signifying an ear and the latter a nose or snout, and is given to this family because the beak is developed at each side into a flat ear-like lobe. The beak is short and stout, and the basal joint of the antennæ reaches beyond the eyes when directed backwards.

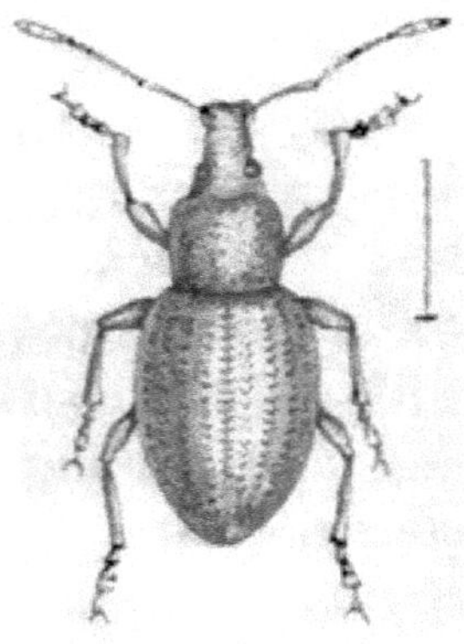

Otiorhynchus picipes.

Of this family our typical example is *Otiorhynchus picipes*, which is here represented. The typical genus, to which this insect belongs, has the antennæ long, and generally set on the tip of the beak. The scutellum is often absent, and where it does exist is very small ;

there are no wings, and the body is egg-shaped and convex. In this genus the ear-like lobes projecting at the tip of the beak, sometimes termed winglets, are very well developed, and can be easily seen with the aid of a magnifier. By these projections there is a deep groove in which the antennæ are set; and in many of the species the head has a most curious resemblance to that of a moose when viewed sideways, the resemblance being increased by the hairs with which the muzzle, if we may so call it, is thickly set.

The insects of this genus are very destructive to plants and fruit-trees, some species attaching themselves more particularly to definite plants, but the generality being in no wise particular as to the sort of plant, tree, or flower on which they feed. The present species is very plentiful, and prefers young leaves to every other kind of food. This fact shows that it is one of the insects that are found in the spring time, and, by beating whitethorn hedges about April and May, any number can be procured. Although not brightly coloured, it is rather a pretty Beetle. Its colour is chestnut-brown, and the elytra are covered with a thick coating of scales, through which a few stiff and shining bristles project. They are boldly striated, each stria being composed of a series of circular impressions, and between the striæ is a row of elevated, smooth, blackish tubercles. The thorax is very globular and thickly granulated.

NEXT comes the family of the Erirhinidæ. This name is compounded from two Greek words, which signify 'long-beaked,' and is given to the family because their beaks are of considerable length, nearly as long indeed as the thorax. The first pair of legs are set very close to each other.

In the accompanying woodcut is seen a magnified representation of Beetle belonging to this family. Its name is *Pissodes pini.* This genus has the beak quite as long as the thorax, and the body is egg-shaped, but long in proportion to its diameter. The species which is represented in the illustration is a northern insect, and is found in Scotland, where it is plentiful among fir-trees. It is a handsome Beetle, its colour being rich red-brown, variegated with golden spots.

Mr. Rye describes the habits of this insect in the following words:—'*Pissodes* . . . frequents pine forests, one species, *Pissodes pini*, abounding in many parts of Scotland, where I have seen the female with her rostrum deeply buried in the soft part between the outer bark and solid timber of fresh-cut fir-trees. In the hole thus formed an egg is deposited, the larva proceeding from which eats galleries under the bark until it is full-grown, when it closes its retreat with particles of wood, grass, &c., and changes to a pupa. The perfect insects . . . cling very tightly to the fingers when handled.' The name *Pissodes* is formed from a Greek word signifying pitch, and is given to these insects because they inhabit the fir-tree.

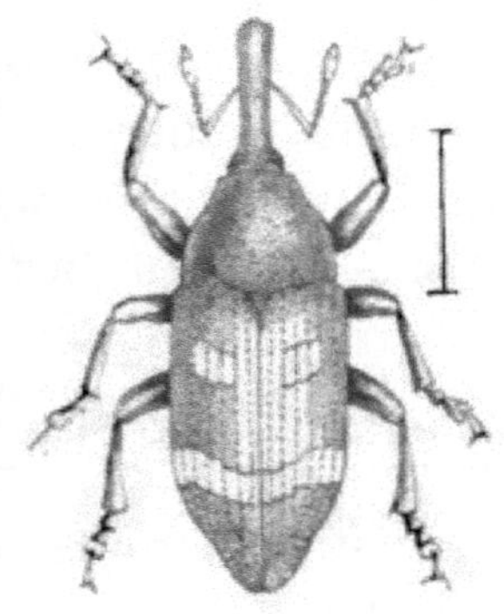
Pissodes pini.

We now come to a Beetle which has doubtlessly annoyed many of my readers, especially if they should happen to be, or to have been, schoolboys. When cracking a filbert after the primitive fashion, it is by no means pleasant to find the shell of the nut yield sooner than expected, and the mouth filled with a bitter black powder, instead of the richly-flavoured kernel. There are few things nastier in their way than such a nut, and the fault lies entirely with the Nut Weevil (*Balaninus nucum*), a figure of which is given herewith. This is a very curious Beetle, its beak being as fine as a needle, very long and very much curved, so that the insect, when viewed in profile, looks something like a shoemaker's awl.

This genus is at once known by the long and slender beak, which is nearly as long as the triangular body. The antennæ are set in the middle of the beak. The present species is rather prettily coloured. The general colour is soft-brown, but the elytra have a nearly white mark shaped like the letter U, its outlines being defined by two black bands. The scutellum is white. These colours are produced by the clothing of down with which the insect is covered, and when the down is rubbed off, the Beetle becomes nearly black.

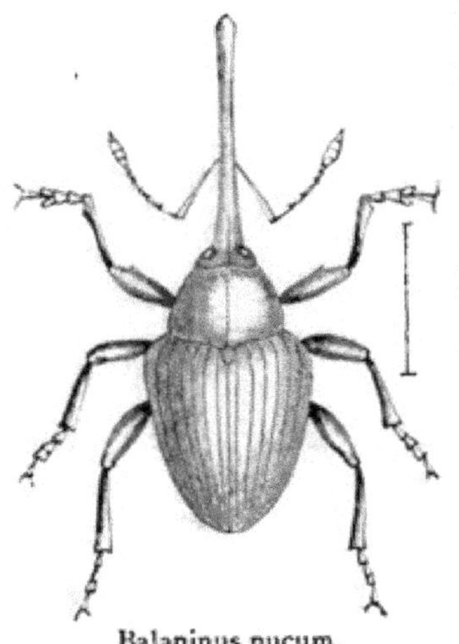

Balaninus nucum.

The life story of this Beetle is very simple. As soon as the nut blossom has fallen, and the fruit has fairly 'set,' the female Weevil begins her work. She bores a hole into the young and still soft fruit, and in the hole she deposits a single egg, repeating the process until she has disposed of her whole stock of eggs. Her business in life is now finished, and she dies. Meanwhile, the eggs are hatched, and the young larvæ begin to feed on the substance of the nut, carefully avoiding the germ, or vital part, so that, to all external appearances, the nut is perfectly sound and good, though three-quarters of its substance may have been eaten by the larva—the little white, fat-bodied grub which we all know so well. As soon as the larva is full-fed, it nibbles a round hole through the shell of the nut, escapes through it and falls to the ground, into which it wriggles its way, and then undergoes its transformations.

As the grub is concealed within the nut until all the mischief is done, there is scarcely any possibility of checking the evil. It has been suggested that, as the nuts which have been attacked by this Beetle become rather loose on their stems, the branches should be beaten before the nuts are ripe, and all the fruit that falls should be burned. Eight species of this genus are known. The name *Balaninus* is derived from a Greek word, signifying an acorn, because the acorn as well as the nut is attacked by species of the same genus.

WE now come to a family of Weevils called Cryptorhynchidæ. This name is composed of two Greek words, signifying 'hidden snout,' and is given to this group of Beetles because they have the beak bent downwards, and capable of being received into a groove on the under side of the body. This attitude is assumed when the Beetle is in repose or alarmed. In the accompanying woodcut is represented one of these insects, named *Cœliodes quercûs.* In this genus the hollow in which the beak lies is between the first and middle pairs of legs, and it is on account of this channel that the name *Cœliodes*, or 'hollowed,' has been given to the genus.

These are all very little, dumpy-bodied, sober-coloured insects, and, when placed under the microscope, they bear a curious resemblance to the apteryx, whose round body and long curved beak almost exactly reproduce the form of the Weevil. They are generally to be found on nettles, and can be taken with the sweep-net. Owing to the rotundity of their bodies, they are very difficult subjects for the setting board. Moreover, in death, the head always bends itself downwards, and the beak tucks itself so firmly into its groove, that to bring it out without injuring the insect is no easy matter.

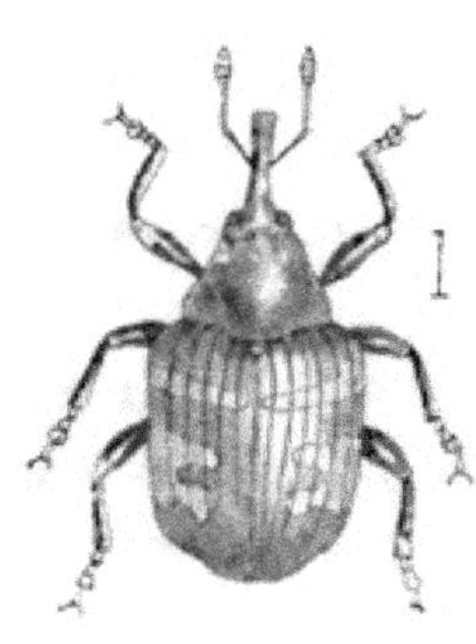

Cœliodes quercûs.

Although this species is not a brilliantly-coloured one, it is very pretty when viewed through a lens. The colour is pitchy-black, the head and thorax being thickly granulated. The elytra are striated and punctated, and are variegated by a few whitish patches. The under surface of the body is clothed with white down. As its specific name implies, it is to be found upon the oak.

Omitting of necessity several genera of these insects, we come to one which is among the most destructive of the group. This is the RICE WEEVIL (*Sitophilus oryzæ*), which is shown in the illustration which accompanies this description. The generic name of *Sitophilus* signifies 'grain-lover,' and is given to the insect on account of the terrible havoc which it makes in corn-stores. It belongs to the family Calandridæ, of which there is only one British genus, that which has just been mentioned. In this family the antennæ have eight joints, the last joint being large and rounded. The body is rather flat, and the elytra, which are boldly striated, do not quite cover the end of the abdomen. There is another species, the CORN WEEVIL (*Sitophilus granarius*), which feeds upon corn as the present species does on rice. The Rice Weevil is distinguished by having four red spots on the elytra, the Corn Weevil being altogether dusky-red.

Like the Nut Weevil, these insects do their destructive work in secret, and there is no finding out the mischief until it is too late. The mother Weevil – tiny herself—bores a tiny hole in a grain of corn, and therein deposits a single egg. The larva is soon hatched from the egg, and sets to work at feeding in the interior of the grain, the whole of which it consumes, leaving the exterior untouched, so that the grain appears quite sound. Of course, the damaged grain is lighter than the sound one, the body of the larva not compensating in weight for the amount of substance devoured by it; and if the corn be thrown into water, the damaged corn will rise to the surface, and may be skimmed off and burned. As, however, damaged corn and rice can both command a sale, and as all damaged grain is not attacked by the Weevil, the dealers will seldom employ such a measure; though to destroy all the light corn for the sake of killing the Weevils would in the long run be more profitable than keeping it for sale and allowing the Weevils to live.

The destruction wrought by these tiny foes can scarcely be over-estimated, but some idea of it may be estimated from the following statements which were made at the Entomological Society, April 4, 1870. Seventy-four tons of Spanish wheat had been carefully sifted or 'screened' to separate the Weevils from it, and out of this quantity *ten hundredweight of Weevils* were sifted. Again, one hundred and forty-five tons of American maize were subjected to the same process, and at two siftings *a ton and three quarters* of Weevils were removed. Now, each of these Beetles had consumed several times its own weight of corn before it attained the perfect state; and the reader may see that, if the grain had been subjected to the water-test and the light portion burned, the proprietor would have saved the cost of some two tons of corn, instead of allowing it to be eaten by these insect devourers, the stock of whom increased in proportion to the diminution of the stores. It is rather a remarkable fact that all these Beetles were Rice, and not Corn Weevils,

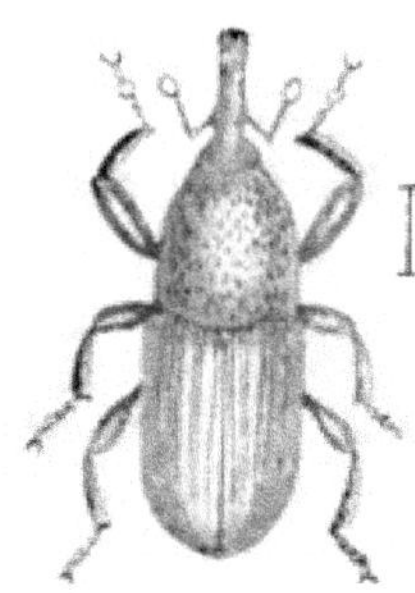

Sitophilus oryzæ.

although there was no rice among the grain which they so seriously damaged.

The larva of these Beetles is very short, fat, and thick, and has two recurved hooks at the end of the body. It remains inside the grain throughout its larval and pupal life.

In the following family of Weevils—namely, the *Hylesinidæ*, or 'wood devourers'—the head has but a very short beak, and is globular in shape and deeply sunk in the thorax. The antennæ are elbowed, and have a long basal joint and a flattened club. The front tibiæ are broad at the tip, and used for digging purposes, and the mandibles are short, strong, sharp, and triangular. They are all timber-feeders, and sometimes work terrible destruction in the forests, even when the trees are still living.

As a typical example of this group we will take the most destructive wood-eating Beetle that we have in this country. Its name is *Scolytus destructor*, and a very appropriate name it is, the generic name being of Greek origin, and referring to the winding passages or burrows which it makes when in the larval state, and the specific name explaining itself. There are six species of British Scolyti, but the present example serves as the best type of the whole genus. This genus is distinguished by the shape of its body, which is obliquely cut off behind, and by the club of the antennæ, which is three-jointed, solid, and flattened. The last joint but one of the tarsus is cleft.

The colour of our species is slightly variable, but is mostly as follows:—The head is black, wrinkled longitudinally, and the thorax is very large in proportion to the size of the insect, and is covered with very small punctures. The elytra are sometimes black, sometimes pitchy, and sometimes bright-chestnut, and are striated, the spaces between the striæ being punctured. So much for the appearance of this Beetle—we will now proceed to its history.

When the mother Scolytus is about to deposit her eggs, she flies to a tree, and searches about the bark for a favourable spot. Having found it she sets to work and gnaws a hole completely through the bark, until she gets between the bark and the solid wood. She next drives a tunnel, scarcely wider than her own body, and then goes back along the tunnel, and deposits her eggs along it. In many cases she exhausts all her life-powers in the effort, and dies before she can entirely escape from the burrow, the entrance of which is stopped up by her body, so that no foe can enter.

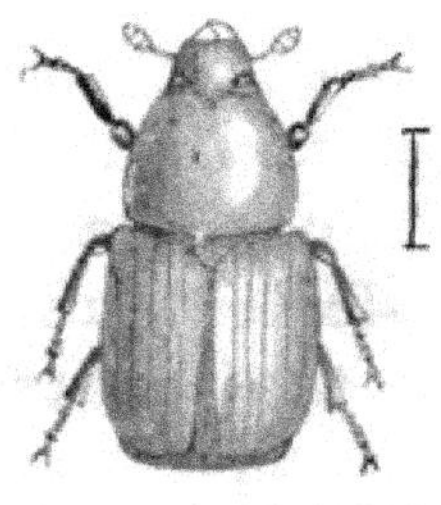

Scolytus destructor.

The eggs are soon hatched, and then the larvæ begin their destructive work. They feed on the soft inner bark, and each larva, as it feeds, instinctively turns itself at right angles to the burrow in which it was hatched, and gnaws for itself a tunnel, which widens in proportion to the growth of the larva. These burrows extend for an inch and a half or two inches in length, and the result is, that a piece of bark, some three inches or more in diameter, is completely severed from the tree, and can no longer perform its office. At the widened end of the burrow the larvæ assume the pupal form, and, after undergoing their change into the perfect insect, gnaw their way through the bark, and are ready to lay the foundations of new colonies.

When a great number of these insects bore into a tree, they often destroy it entirely, the bark being separated as completely from the wood as turf is severed from the ground when the spade is passed under it. There is a tree—or rather, what was a tree—standing within a few yards of my house, which has been stripped by the Scolytus. The whole of the bark has peeled off, and nothing is left but the naked wood, scored all over with the radiating tunnels of the destroying insect.

It is much doubted whether the Scolytus ever attacks a healthy tree, principally, as is conjectured, because in such trees the burrows of the insects are filled with sap, which not only drives out the Beetles, but prevents their eggs from being hatched. Still when a tree becomes unhealthy, the attacks of the Scolytus prevent it from recovering itself; and such serious damage has been done by this insect to our trees, especially the elms in and about London, that the attention of entomologists has long been directed to the subject, in hopes of discovering some device by which the ravages of the Scolytus may be checked, if not altogether stopped. As yet, however, no scheme has succeeded. Various plans have been suggested, such as injecting poisonous fluids into the hole made by the mother Scolytus. This might possibly answer, provided the operater could be sure of discovering all the holes, and provided that the liquid did not kill the tree as well as the insect. The 'Gishurst Compound' would do as much in this way as anything could, but it cannot be employed on a large scale.

At present, the opinion seems to be that the only plan which offers the least probability of success is a 'stamping out' process, similar to that which saved us in the time of the cattle-plague. By this plan, all trees which are visibly attacked by the Scolytus are to be cut down, and stripped of their bark and the outer layer of wood, which are then to be burned, so as to destroy the Scolytus, its larvæ, pupæ, and eggs.

CHAPTER XII.

LONGICORNES, OR LONG HORNED BEETLES.

THIS group of Beetles derives its name from the shape of the antennæ, which are generally long, though in some of our commonest species they are only of moderate length; but, whether they be long or short, they are never clubbed, and are mostly slender and thread-like. Their head is not lengthened into a beak like that of the preceding group, and the elytra are always broader at the base than the thorax. There are other peculiarities of structure, but these are quite sufficient to distinguish them. Indeed, there is something so characteristic in the appearance of a Longicorn Beetle, that even a novice finds no difficulty in recognising it.

They are all wood-borers in the larval condition, and are thin, long, whitish grubs, rather flattened, and with the segments boldly marked. By means of this latter structure, the larvæ are able to force their way through the wooden tunnels in which they live. They possess the usual six legs, but these limbs are only rudimentary, and of no use in locomotion.

As the grub has to feed upon hard material, it is furnished with very strong horny jaws, and, in order to accommodate the muscles which move these jaws, the head is very broad and covered with a hard skin, nearly as strong indeed as the jaws themselves. In consequence of their habits, the proceedings of the larvæ are difficult of observation, and require machinery such as few entomologists can hope to possess. Still, by carefully opening the trees which are infested by these destructive insects, much can be learned of their habits, and many pleasant and instructive hours can be spent in this task. We will now proceed to examine some of the British species of the Longicorn Beetles.

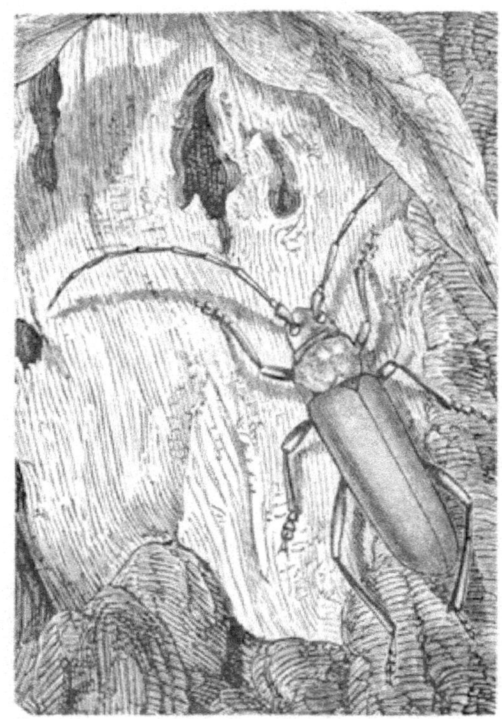

Aromia moschata.

OUR first example is the beautiful MUSK BEETLE (*Aromia* [or *Cerambyx*] *moschata*), an insect which is equally grateful to the eye and the nostril.

This is really a lovely insect, the shape being peculiarly elegant, and the colour a soft green, sometimes glossed with blue, gold, copper, or bronze, the hue being exceedingly variable. Even the thorax partakes of this variability, being sometimes rough, and sometimes very smooth and glossy.

It owes its popular name to the powerful and pleasing odour which it exhales, and which is said by some persons to resemble ottar of roses. For my part, I never could perceive much resemblance between the somewhat sickly scent of the ottar and the fresh sweetbriar-like odour of the Beetle. This scent can be perceived at a considerable distance, and the presence of the Beetle can often be detected by it, even when the insect is itself concealed. Shortly after I came to live in West Kent, I was passing along the road, and declared that a Musk Beetle was in the neighbourhood. My companions, not knowing the scent, could not believe me, and made themselves rather merry on the subject. In the course of the day, however, one of the same party, a young lady, was passing by the same place, and carried off a fine Musk Beetle in her hair.

The scent of this insect is as enduring as it is powerful, and, if the Beetle be held with a gloved hand, or wrapped in a handkerchief, it will impart either to the kid or cambric its peculiar odour, which will last for a very long time. From a series of experiments made some few years ago, I have come to the conclusion that the Musk Beetle can emit or retain its odour at pleasure as long as it is in full health, but that when the insect is weak, or in a dying state, it is unable to retain the scent.

Wherever willow-trees are to be found abundantly, there the Musk Beetle is sure to be, because it feeds on the interior of that tree while in the larval state.

The ground on which my house stands being very high, and the soil being gravel, I was very much surprised at perceiving the Musk Beetle which has just been mentioned, thinking that no willow-trees were near. However, after a while, I came upon some of these trees, at a distance of some 300 yards, growing on the banks of a little stream that ran in the valley below. Sometimes a tree is absolutely riddled with the burrows of these larvæ, which bore deeply into the very heart of the timber, and leave little except a shell of bark surrounding a sort of soft wooden sponge.

Those who wish to capture the Musk Beetle will find that they cannot do better than explore the largest, the oldest, and the most rugged willow-trees. The Musk Beetle is not a very active insect, and is fond of clinging to the bark of the willow, and remaining perfectly still for many hours together. I may mention that the surface of this Beetle affords a most gorgeous object for the microscope.

Clytus arietis.

OUR next example of the *Longicornes* is a very common and very pretty Beetle, and known to entomologists by the name of *Clytus arietis*. In this genus the antennæ are shorter than the body, the last joint being somewhat conical. The palpi are short, with the last joint stout and three-sided, the angles being rounded. The thorax is globular, and the body cylindrical.

The colour of this species is black, with three yellow bands across the elytra, and a yellow patch at their tips, so that the insect has a very waspish look, and is popularly known as the WASP BEETLE. The similitude is increased by its fussy mode of walking, and the perpetual movement of its antennæ, and, as it crawls in and out of the foliage on hedges, it has so very wasp-like a look that few persons, except they be entomologists, like to touch it. In its larval state it burrows into wood, and emerges somewhere about midsummer. It is fond of frequenting flowers, and can be taken in almost any quantity. Being rather variable in the hue of its markings as well as in size, a series ought to be taken for the cabinet.

THE last Longicorn Beetle which space will allow us to describe is known by the name of *Strangalia armata.* In this genus the thorax is without spines, is narrow in front and flattened above. The body is very narrow and almost pointed behind. The front of the head is rather lengthened. The present species has the ends of the elytra deeply cut, so as to form a rounded notch with toothed edges, and the male can be recognised by two conspicuous tooth-like processes on the inner side of the hinder tibiæ. Owing to these peculiarities, the specific name of *armata*, or 'armed,' has been given to the Beetle. In colour and

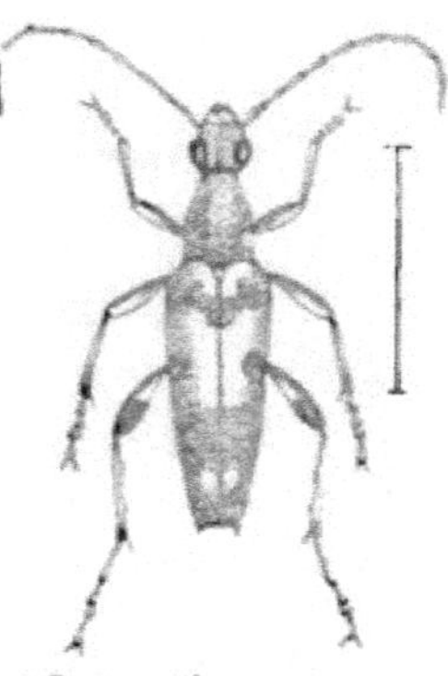

Strangalia armata.

size it is one of the most variable of British Beetles, scarcely any two specimens being exactly alike. Sometimes it is almost entirely black barred with yellow, sometimes yellow barred with black, while it varies in length from five lines to three-quarters of an inch. Being so variable, the entomologist ought to have a series of specimens in his cabinet. There is not the least difficulty in obtaining them, as it is a very common Beetle, and can be found in plenty upon the umbelliferous plants—these being the favourite resorts of many Beetles. It is tolerably active, and takes to the wing as soon as it is alarmed. Seven species of this genus are known in England.

CHAPTER XIII.

EUPODA.

WE now leave the Longicorn Beetles, and come to another section, called the EUPODA, a name derived from two Greek words, signifying 'beautiful feet.' By some authors the section is named *Phytophaga*, this word being also of Greek origin, and signifying 'plant-eater.' They are all pretty insects, and some of them, though not large, are singularly beautiful both in form and colour. In the Beetles belonging to this section the antennæ are short and slender, and have a short basal joint. The head is deeply sunk into the thorax, the elytra cover the sides of the abdomen, and the last joint but one of the tarsus has two lobes. They are all vegetable-feeders, and, as most of them are attached to certain plants, there is little difficulty in finding them.

Passing by the family of the Sagridæ, of which we have but four species in England, all belonging to one genus, we begin with the Donaciadæ. In this family the antennæ are longer than is usually the case with the Eupoda, and they are set just in front of the eyes. The head is large in front and narrowed behind, and the first segment of the abdomen is very long. Our example of this family belongs to the typical genus, and is known to entomologists by the name of *Donacia menyanthidis*, a figure of which is given in the accompanying woodcut. In this genus the body is flattened, polished, and shining above, thickly punctured, and having altogether a metallic aspect. Below, it is covered with a very fine down. The antennæ have the fourth and following joints elongated.

All the Donaciæ may be found on water-plants, especially on reeds, from which they derive the generic name, Donax being a Greek word, signifying a reed. Although some of the species are rather rare, the Donaciæ are, beyond comparison, the most common of water-frequenting Beetles, and the leaves of reeds, water-lilies, and other plants are often studded with these beautiful insects, whose polished and variously-coloured bodies glitter in the sunbeams like living gems. As many of the species are exceedingly variable in colour, it will be as well for the entomologist to procure a considerable number of specimens, many, which at first sight appear to be different species, being, when closely examined, seen to be nothing but varieties of the same species. There is scarcely a colour of the rainbow which is not exhibited by one or other of the Donaciæ, and in some instances the same species exhibits an astonishing

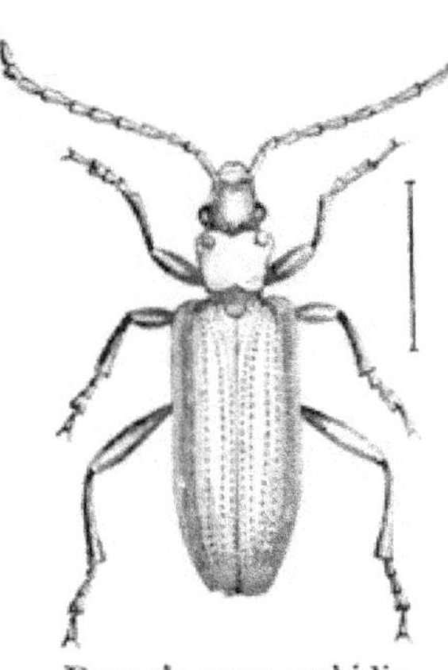
Donacia menyanthidis.

variety of colour—one being perhaps almost black, another blue, another green, another copper, another purple, another red, while another is as made of burnished gold.

The larvæ of these Beetles live within the stems of the various water-plants, and on that account the species have derived their names from the plants on which they live.

Altogether, exclusive of varieties, of which there are a great number, nineteen British species of this beautiful genus are known. In some places these Beetles are so plentiful that seven species have been found on water-plants within the limits of one small pond.

The present species is shining-green above with a brassy gloss, and below it is silvery-white, owing to the soft down with which it is clothed. There is a bold groove on the front of the head, and the elytra are both striated and 'crenated,' i.e. covered with little marks formed like segments of circles. It is not a very common species, being usually confined to certain localities.

THE best known species of the genus *Crioceris* is the ASPARAGUS BEETLE (*C. asparagi*), which feeds on the plant from which it derives its name. This insect is much longer in the body than the preceding species, though smaller in point of bulk. It is very prettily coloured, the thorax being deep-red, and the head and elytra shining-blue or green-black, the latter being marked with reddish-yellow, so as to look as if they were yellow on which a black cross had been laid. The larvæ are grey, soft-bodied creatures, and, together with the perfect insect, can be found in any number upon the asparagus after it has been allowed to run to seed, and wave its feathery branches and pretty round fruit in the air.

NOW we come to a family in which most of the species are brilliantly coloured; and even those in which the hue is apparently of a sombre cast are seen, when closely examined, to be really clothed with as much beauty as their more conspicuous relatives. This family is called the Chrysomelidæ, a name which is composed of two Greek words, signifying 'golden apple,' and is appropriately given to these Beetles on account of the globular shape of their bodies, and the lovely tints with which they are adorned. In these Beetles the head is very far sunk in the thorax, but, not so deeply as in the last-mentioned family, and the antennæ are stouter, shorter, and more thickened towards the tip. The body is oval or round, and the legs are of equal size. In the genus Timarcha, from which our example is taken, the wings are not developed, and the elytra are firmly soldered together at the suture, so that they cannot be opened.

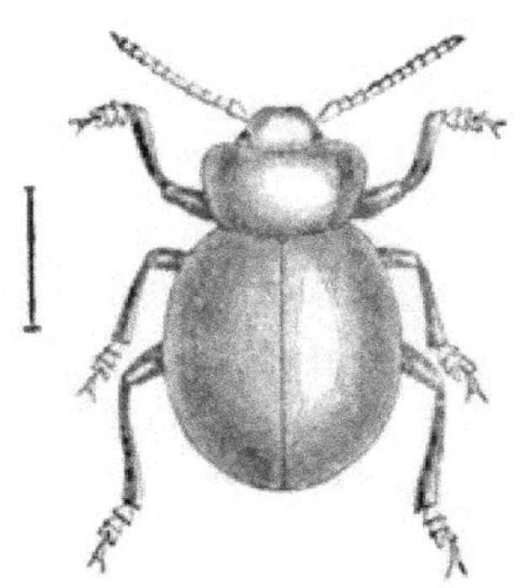

Timarcha lævigata.

Herewith is represented an insect which is very plentiful, and known by the name of *Timarcha lævigata*. It is better known, however, by the popular name of BLOODY-NOSE BEETLE, because it has a habit of ejecting a large drop of red fluid from its mouth when it is handled. There are only two British species belonging to this genus, and they are by far the largest English representatives of the Chrysomelidæ. The present species sometimes nearly reaches a length of three-quarters of an inch, and, as the body is very stout and globular, it may take rank among the larger British Beetles.

At the first glance, this insect appears to be black, but a careful examination with a magnifying glass, aided by a strong light, shows that the real colour is the deepest indigo-purple, sometimes with a gloss of green. The whole of the upper surface is thickly covered with small punctures, which impart to it a sort of velvety gloss. The tarsi are very broad, and furnished beneath with thick, greyish-yellow pads. There is a very broad impression on the forehead. The second species, *Timarcha coriacea*, much resembles the present insect. It is not, however, so large, and is rather blacker above. The chief distinction, however, is to be found in the punctures of the elytra, which are deep, and have a tendency to run together, so as to form irregular lines. This formation cannot be seen without the use of a magnifying glass.

The larva of this Bloody-Nose Beetle is wonderfully like the perfect insect. It is large, heavy, soft-bodied, and covered with a shining blue or green skin, a yellowish patch appearing at the apex of the body. It is very common in the ditches under hedgerows, especially if the hedge be allowed to flourish in the luxuriant and picturesque manner which is so fascinating to an artistic eye, and so hateful to the agricultural eye of the farmer, who cares nothing for beauty, and would sacrifice the loveliest country scene in England to get another cart-load of turnips out of his field. How these great, sluggish, conspicuous larvæ ever escape the many perils of larval life is really wonderful. I can only account for their survival on the supposition that they are distasteful to the insect-eating birds. Very many larvæ of this section secrete a bitter, or acrid liquid, and the Timarcha larva may perhaps be protected by some such means.

IN the accompanying illustration will be seen depicted a member of the typical genus, *Chrysomela staphylea*. In this genus the wings are fully developed, and the last joint of the palpi is rather hatchet-shaped. The present species is a moderately large one, and has the body extremely convex. The general colour is reddish-brown with a slight metallic gloss. The whole upper surface is covered with punctures, those of the head and thorax being very fine, and those of the elytra large and irregularly disposed. Beneath, the body is pale-brown. There are about twenty British species of this beautiful genus. The present species takes its name from the common Bladder-nut (*Staphylea pinnata*), upon which it can be found. It is a very common insect.

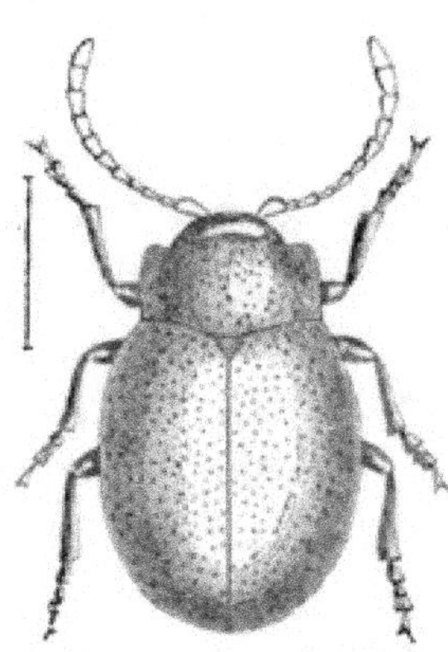

Chrysomela staphylea.

PASSING by the family of the Galerucidæ, we come

to the enormous family of the Halticidæ, the members of which, though individually small and insignificant, collectively exercise very great influence upon the agriculture of our country. One of these insects is shown beneath, and is one of several Beetles which are known by the popular name of TURNIP-FLEAS or HOPPERS. It is called scientifically *Phyllotreta* (or *Haltica*) *brassicæ.*

All the Halticidæ can be distinguished by the very thick hinder femora, which denote the possession of great leaping powers on the part of the insect. The antennæ are set between the eyes, and the edges of the elytra are wavy. They are attached to different plants, and are so constant to them that their specific names are often taken from their food-plant. The colour of the present species is deep-black, and there are two longitudinal yellow streaks upon each of the elytra, one near the base, and the other towards the apex. The body is egg-shaped, and the elytra are rounded at their tips. It is about the smallest species of the genus.

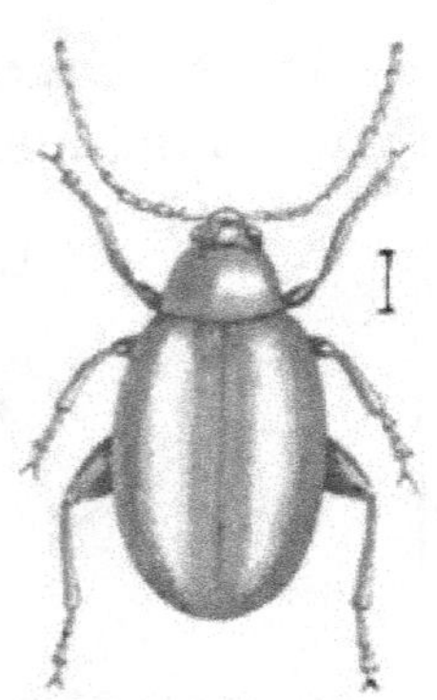

Phyllotreta brassicæ.

The true Turnip-flea is another species, called *Phyllotreta nemorum*, the colour of which is black, with a broad longitudinal yellow streak running nearly, but not quite, to the end of the elytra. Both in the larval and perfect stages this insect is extremely damaging to the turnips, feeding upon the first tender shoots as they appear above the ground, and so destroying the whole plant. Sometimes they will devastate a field so completely, that it must be re-sown if a crop be wanted.

The name *Phyllotreta* is of Greek origin, and signifies 'leaf-borer.' There are more than a hundred British species of these little Beetles, and the now restricted genus *Phyllotreta* contains fourteen species.

We now pass to the family of the Cassidiidæ, commonly known by the popular and appropriate name of TORTOISE BEETLES. There is no possibility of mistaking these Beetles, which are flat-bodied, rounded, and have the head completely hidden under the wide thorax, which overlaps the base of the elytra. As the insects sit upon leaves, the whole of the head and limbs are completely hidden by the thorax and elytra, just as are those of a tortoise by its shell. Most of the species are green, though some of them are adorned with spots and stripes of red and gold. All these colours are exceedingly fugitive, and vanish soon after the death of the insect. Glycerine has been tried with some of these insects, but with little, or no success.

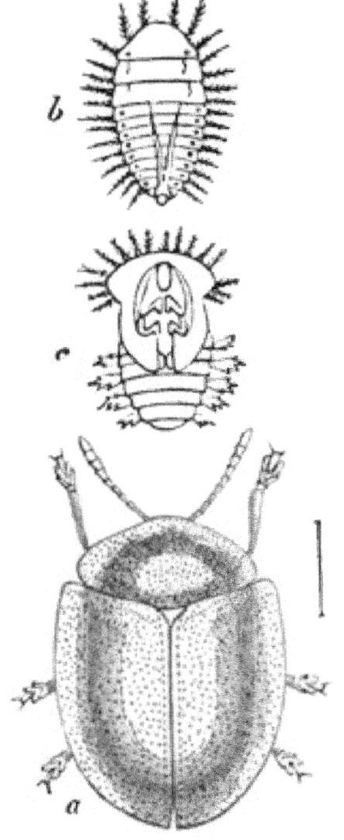

Cassida viridis.

A very common species, *Cassida viridis*, is shown herewith. It is of a rather dull-green colour, the base of the elytra taking a slightly reddish hue. It may be found plentifully on thistles.

At Fig. *b* of the same illustration is shown the extraordinary larva of the Tortoise Beetle. It is very flat, and has the sides covered with flattened spines. But the most curious portion of its structure is the forked apparatus that proceeds from its tail and passes over its back. This fork serves a very curious purpose. Like several other larvæ, both carnivorous and vegetable feeders, this creature is sheltered by a sort of umbrella formed of the refuse of its food. But, instead of lying directly on the back, the umbrella is supported on the fork at some little distance from the body, and, when it becomes too weighty it can be shaken off and a new one gradually produced. The pupa is scarcely less curious in appearance, and is drawn at Fig. *c*.

There is only one British genus of this family, and it contains about thirteen species.

CHAPTER XIV.

PSEUDOTRIMERA.

WE now come to the last section of the Beetles, the PSEUDOTRIMERA. This name is compounded of three Greek words signifying 'false three-joints,' and is given to the insects on account of the structure of their tarsi, which appear to have only three joints, though in reality they have four joints; the missing joint, which is the third, being very minute, and hidden within the doubly-lobed second joint.

THE first family upon our list, namely, the *Coccinellidæ*, is composed of insects which are very familiar to us under the popular name of Lady-birds or Lady-cows—the former being the more common as well as the more poetical name. These insects are all flat below and convex above; the body is rounded and the antennæ are short. The typical genus, *Coccinella*, has the hinder angles of the thorax acute. Many species of these pretty insects inhabit England, but some of the species are so exceedingly variable in point of colour, that the varieties have been described as actual species by practised entomologists. It is common enough to find a blue insect running into shades of green, and *vice versâ*; but in some of these Lady-birds, the same species will be red spotted with black, black spotted or blotched with red, black spotted with yellow, yellow spotted and barred or blotched with black, and so on in infinite variety.

We can briefly describe but one species, a very common one, and then proceed to the habits of the insect.

The best-known of the British Lady-birds is the common SEVEN-SPOT LADY-BIRD (*Coccinella septempunctata*), a species that derives its name from the seven black spots upon the elytra. These spots are exceedingly variable in size, and sometimes one or two spots are absent, while there is one variety in which there are no spots at all. All the Lady-birds possess long and ample wings, as they have to fly to considerable distances.

Beautiful as are the Lady-birds, it is not for their beauty alone that they are valued, inasmuch as they are among the greatest benefactors of civilised man, and preserve many a harvest which, but for their aid, would be hopelessly lost. For, in their larval state they feed upon the aphides—the 'green blight' or 'green-fly' of gardeners—and, being exceedingly voracious, devour vast numbers of those destructive insects. Few persons would suppose, on looking at the Coccinella larva, what was its real condition of life. It looks as harmless, dull, sluggish a creature as can be imagined, and much more likely to be eaten itself than to eat other insects. Yet, with all this innocence of aspect, it is so ruthless a destroyer of animal life, that if a few of them be placed on a bush or plant which is infested with aphides, in a day or two not an aphis will be left.

Especially is the Lady-bird useful in those parts of the country where hops are grown. There is an aphis which feeds especially upon this plant, and which has been known to destroy whole plantations in a single season, causing the greatest distress among the multitudes whose living depends more or less directly upon the hop. Fortunately, the Lady-bird—usually the little TWO-SPOT LADY-BIRD (*Coccinella bipunctata*)—comes to the rescue, and follows the aphides wherever they are most plentiful. The mother insect lays her eggs in packets among the aphides, and, as soon as the young larvæ are able to move about, they begin to feed upon the insects near where they have been placed.

In some seasons the swarms of Lady-birds almost exceed belief. I have seen the streets absolutely red with them, and the houses covered with their multitudes, while within doors a thick band of Lady-birds ran along the angle of the walls and ceiling like a red cable, large bunches hanging in each corner. These insects very well illustrated the adage that 'dirt is only matter in the wrong place.' Nothing could be more beneficial than their presence in the locality, as it was situated in the very midst of hop gardens, and by their means the year's harvest was saved from destruction. But, though they were much wanted out of doors, they were not at all wanted inside the house, especially as Lady-birds have a very unpleasant odour, which, when multiplied by tens of thousands, becomes almost unbearable. Even after the rooms had been cleared, they were almost uninhabitable, and the more so that it was impossible to keep the windows open, because the Lady-birds flocked into the room in swarms, and would soon have replaced those which had been ejected. Even throughout the winter many of them retained their positions, having been kept alive by the warmth of the fire.

When the larva is full-fed, it attaches itself to a twig or leaf by the end of its tail, and thus hangs with its head downwards. Presently, the larval skin splits down the back, but the pupa does not emerge, remaining within the larval skin until it has changed into its perfect form. It has been mentioned that the Lady-birds give out a very unpleasant odour. This is caused by a yellowish liquid which issues from the joints of the limbs, as has been described in connection with the Oil Beetle, and which has a very powerful and disagreeable scent. In some parts of the country this liquid is considered to be a cure for toothache, the finger being first rubbed against the legs of the Lady-bird, and then on the offending tooth. In its larval state it emits a similar liquid from the tubercles upon its body.

THE family of the Trichopterygidæ contains a good many species, of which we select one as our example. This is *Trichopteryx atomaria*, which is represented on the next page.

These are all little Beetles, and, indeed, are the tiniest of the British Coleoptera. Small as they are, they can be easily recognised when examined by the aid of a lens, so bold are the characteristics which mark them. The antennæ are long, slender, and beset with long hairs, and having a bold three-jointed club. The wings are very long and narrow, and fringed with hairs, a peculiarity which has gained for them the name of Trichopteryx, or 'hairy wings.' Sometimes the wings are undeveloped, but when they are present they are always fringed with hair. There are other characteristics of the family, but these are sufficient for the recognition of any insect that belongs to it.

In the typical genus, Trichopteryx, the antennæ are about half as long as the body, the head is convex, large and triangular, and the wings are furnished at their tips with several bundles of hairs. The present species is one of the largest of the family, and yet it is only one twenty-fourth of an inch in length. Small as it is, by the side of other species of the same genus it is really a giant, most being the thirty-sixth part of an inch in length, while there are some which are barely one-hundredth of an inch long. Some notion of the size of these tiny creatures may be obtained by looking at the little line on the right hand of the figure, and reflecting that they measure just one quarter of that length.

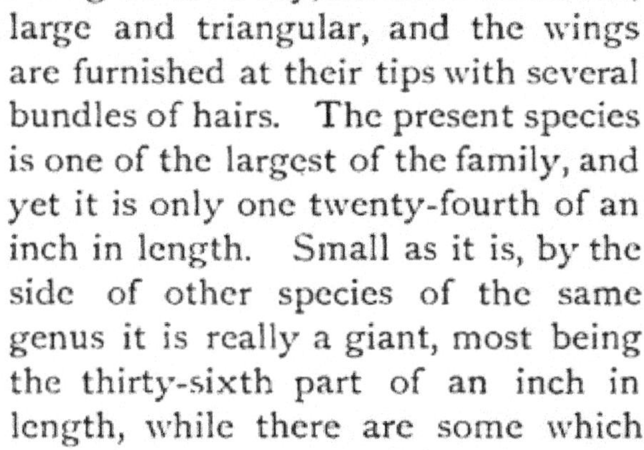

Trichopteryx atomaria.

The little insect which has been chosen as our example of these 'micro-coleoptera,' as the tiny Beetles are called, is tolerably common, and can be found under heaps of decaying leaf-mould and similar localities. Though the finder may not be able to recognise the precise species when he discovers it, he can at all events see that it *is* a Beetle, whereas, when he finds the exceedingly minute creatures which have just been mentioned, it is impossible for him to know, without the aid of a lens, that the little black speck is even an insect, much less whether it be a Beetle or not. The best mode of capturing these Beetles is to take some leaf-mould from under a heap, scatter it thinly on a sheet of white paper, and then go over it carefully with a tolerably powerful lens.

Owing to the very minute dimensions of these Beetles, the exact definition of the species is a very difficult business, but it is evident that many species are known in this country.

PASSING by one family of this section, we come to the Pselaphidæ, of which remarkable family two examples will be given, each illustrating one of the sub-families.

In these Beetles the elytra are very short, so short, indeed, that for many years these insects were classed among the Brachelytra. The club of the antennæ is bold and well-defined, the last joint being very large. The head is narrowed behind into a distinct neck.

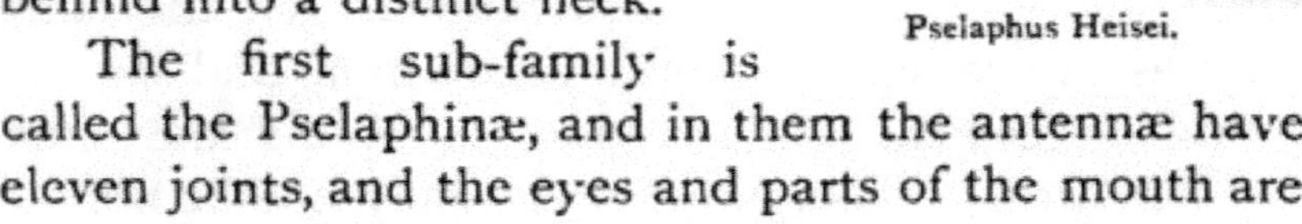

Pselaphus Heisei.

The first sub-family is called the Pselaphinæ, and in them the antennæ have eleven joints, and the eyes and parts of the mouth are well developed. The genus Pselaphus, of which there are only two British species, has the antennæ, palpi, and legs very long. The commonest species, *Pselaphus Heisei*, which is represented herewith, is shining yellow-brown, has its body very flat and wide, and on each side of the suture of the elytra there is a stria which runs from the base to the tip. It can be shaken out of moss, as can its very rare congener *Pselaphus Dresdenensis*, which may be distinguished by its dark colour and a semicircular impression at the base of the thorax.

THE last example of the British Beetles is, perhaps, the strangest of all our native insects, and how it can find any gratification in existence is not easy to see. We feel that the life of a blind and deaf man is a hard one, shut out as he is from free intercourse with his fellow-creatures, and incapable of enjoying, or even of comprehending, the common blessings of sight and hearing. Yet he is capable of one kind of animal enjoyment, for he can eat, and indeed upon this capability is based the course of instruction by which such afflicted persons have been rescued from their wretched isolation, and taught to interchange ideas with their fellow-men. But, supposing that a man who was incapable of sight or hearing were also found without a mouth, and yet possessing the power of living without food, we should think that such a being must have reached the very abyss of misery—a misery beyond all power of alleviation.

Yet in the Beetle which is shown below we see an insect in which these imaginary privations are the normal state, and which possesses neither eyes nor mouth, and is capable of supporting existence without food. We should, however, be very wrong in supposing that this insect must be miserable because a human being under such conditions would be supremely wretched, and may be sure that, in some mysterious way, this Beetle, which leads a darkling life and is incapable of eating, is just as happy in its way as the brilliant butterfly that basks in the sunshine, and flits from flower to flower, enjoying their lovely colours and sweet juices.

Whether this insect be possessed of some senses unknown to us, must of necessity be a problem not likely to be solved, but, as far as we can judge, the only sense which it can possess is that of touch. The name Pselaphidæ refers to this supposition, and is formed from a Greek word, signifying the groping movements of one who tries to find his way in the dark.

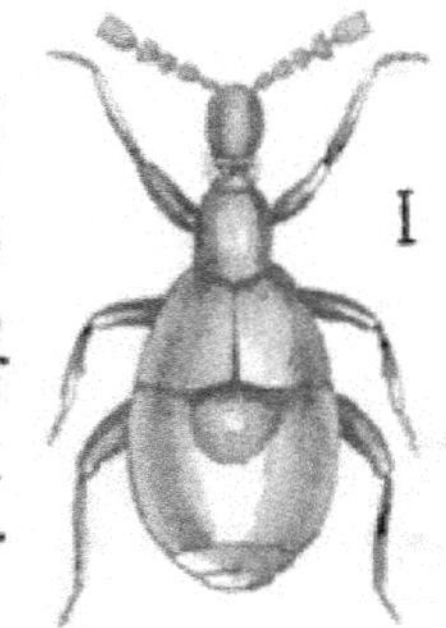

Claviger foveolatus.

The name of this Beetle is *Claviger foveolatus*, the former name signifying 'club-bearer,' and given to the insect on account of the form of the antenna, which is boldly clubbed, and has only five joints. This Beetle can be found in the nests of the yellow ant (*Formica flava*), a very common insect, which makes its nests under large stones if it can find them, or, in default of such shelter, throws up little mounds of earth. It mostly prefers heaths and hilly districts. The colour of the beetle is yellow, like that of the ant with which it lives, and it has no wings. The name *foveolatus* is given to it on account of the deep fovea, or hollow in the middle of the abdomen.

LEPIDOPTERA.

LEPIDOPTERA.

CHAPTER I.

RHOPALOCERA, OR BUTTERFLIES.

I VERY much regret the necessity for using such words as that which appears at the head of this chapter. I employ such words as seldom as possible, and always explain them when compelled to use them, as is the case at present. Still, in many instances, scientific terms are absolutely necessary, because there are no existing English words which have the same signification, and in many others, even though there may be English equivalents, the scientific terms are so universally employed that it is necessary to introduce them and explain their meaning.

To begin with the word Lepidoptera. It is formed from two Greek words, the one signifying a scale, and the other a wing, and is given to those insects because their wings are, for the most part, covered on both sides with flat scales which overlap each other just like the tiles of a house. This is the most important characteristic of the order, but there are one or two others which must be noticed. The mouth is formed for suction. The mandibles, or jaws, which are so conspicuous in the insects which we have hitherto examined, and which indeed are large and powerful in the larval state, are scarcely visible, being reduced to mere rudiments of jaws. The maxillæ, on the contrary, are very much elongated, and modified into the beautiful proboscis through which the insect is able to suck the sweet juices of flowers. The pupa is enclosed in a hard, shelly case, not resembling the perfect insect, this form being scientifically called 'obtected.'

The Lepidoptera are usually supposed to fall naturally into two great divisions or tribes, known popularly as Butterflies and Moths. The distinguishing features of the former may be briefly summed up as follows. The antennæ are terminated by a kind of knob, or club, the body is pinched in at the junction of the thorax with the abdomen, while the hinder wings are rigid and incapable of being folded. There are also one or two minor distinctions, which, however, need not be here mentioned.

With the British Lepidoptera this division into these two great tribes answers well enough, although in both groups there are one or two insects which present rather anomalous characteristics. But, when we come to inquire into the structure of the foreign representatives of the order, we find that this arrangement no longer holds good, for there is no single distinguishing feature of the constituents of the one group which is not possessed by some of the members of the other. Thus the clubbed antennæ of the butter-

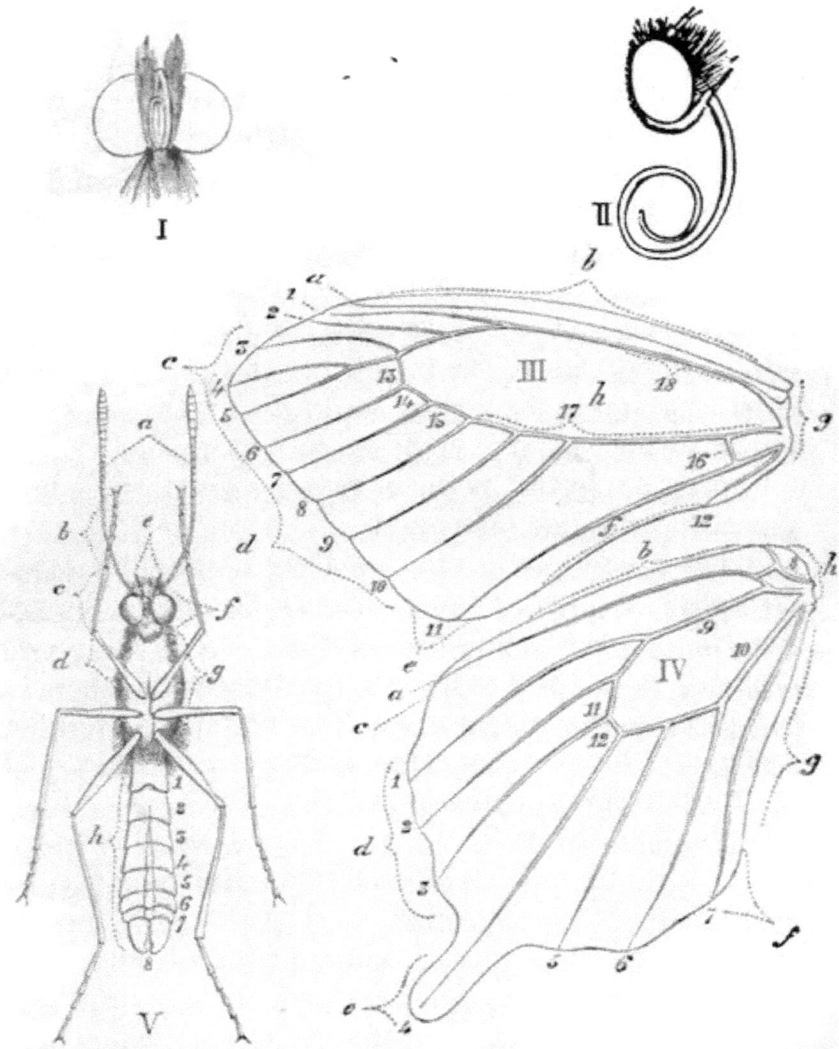

I. *Front view of head.* II. *Side view.*

III. *Fore Wing.* 1—5. Subcostal nervules. **6, 7.** Discoidal nervules. 8—10. Median nervules. 11. Submedian nervure. **12.** Internal nervure. 13—15. Disco-cellular nervules. 16. Interno-median nervule. 17. Median nervure. 18. Subcostal nervure. *a.* Costal nervure. *b.* Costa or anterior margin. *c.* Apex or anterior angle. *d.* Posterior or hind margin. *e.* Posterior or anal angle. *f.* Interior or inner margin. *g.* Base. *h.* Discoidal cell.

IV. *Hind wing.* 1, 2. Subcostal nervules. 3. Discoidal nervule. 4—6. Median nervules. 7. Submedian nervure. 8. Precostal nervure. 9. Sub-costal nervure. 10. Median nervure. 11, 12. Disco-cellular nervules. *a.* Costal nervure. *b.* Costa or anterior margin. *c.* Apex or anterior angle. *d.* Hind margin. *e.* Tail or caudal appendage. *f.* Anal angle. *g.* Abdominal or inner margin. *h.* Base.

V. *Underside of body.* 1—7. Abdominal segments. 8. Caudal or anal extremity. *a.* Antennæ. *b.* Tarsus. *c.* Tibia. *d.* Femur. *e.* Palpi *f.* The head. *g.* The thorax. *h.* Abdomen.

flies are also found in a marked degree in many of the moths, while the slender and contracted body, and the structure of the hinder wings are found in insects of both tribes.

However, as this work treats of entomology in a popular rather than a purely scientific sense, we will accept the more general system of classification, and still retain the title of butterflies and moths, as applied to the insects of the two so-called divisions of the Lepidoptera.

On the preceding page is a map or chart of a Butterfly, showing the principal portions of the insect, and the distinctive names attached to them by entomologists. Some of these words look rather formidable, but there is really little difficulty in learning and retaining them; and the best way of mastering them is, to trace them out on the wings of various Butterflies, and, if possible, to sketch those wings on an enlarged scale, and write the names of the different portions. The principal portions of the wings are those which are denoted by letters, and which should therefore be learned first; the knowledge of the other portions being gained by degrees. Here I may mention that Mr. E. Newman repudiates the word nervure, and substitutes the simpler word 'ray,' as analogous with the fin-rays of fishes. He is undoubtedly justified in considering that 'ray' is the better word, but as in all scientific accounts of the Lepidoptera the word 'nervure' is used, I have employed it, leaving the reader to substitute the word 'ray' if he should prefer it.

The reader will notice the enormous size of the eye-masses, as shown in Fig. 1, this great size and bold projection being rendered necessary by the fact that these insects are all day fliers, perpetually on the wing, and, consequently, very conspicuous. Now, there are many creatures—certain birds, for example, dragon-flies, and other predacious foes—which are very fond of Butterflies, and would wofully thin their numbers did not their multitudinous eyes enable them to see the approaching enemy in time for their broad wings to carry them out of danger. The form of the proboscis is also shown in Figs. 1 and 2, the former representing it as it appears when coiled up so as to be out of harm's way, and the latter showing it as partly uncoiled, as it appears when the insect is about to take food.

WE now proceed to take in their order some typical examples of British Butterflies. The first family is called Papilionidæ, and may be distinguished by having the first pair of legs formed for walking, the tip of the antennæ not hooked, and the discoidal cell of the hind wings quite closed. Only one genus inhabits this country, and only one species, the beautiful SWALLOW-TAIL Butterfly (*Papilio Machaon*), which is shown in the frontispiece.

In the genus to which this Butterfly belongs, the hind wings are tailed, and the caterpillar, or larva, is furnished with a forked appendage called the 'nuchal horn' because it issues from the neck. In this species the horn is only used in moments of irritation, and is concealed within the body, its place being only marked by two dots. If, however, the caterpillar be irritated or hurt, it immediately throws out the horn, which can be protruded to the length of half an inch or so. Many naturalists suppose that this horn is intended for the purpose of driving away the ichneumon flies when they attack the larva. I can, however, scarcely accept this theory, because the ichneumon flies are terribly injurious to many other caterpillars, which yet are supplied with no apparatus for driving them away. This organ, whatever purpose it may subserve, gives forth a very strong odour, much resembling that of fennel, and so powerful that even in the open air it can be perceived at some distance.

The colour of this splendid Butterfly is almost entirely yellow and black. On the lower wings, however, there is a row of six cloudy blue spots, sprinkled with yellow dots, and at the anal angle of each lower wing is a large red spot with a slight blue crescent on the upper part. This Butterfly was once spread over a considerable part of England, but now seems to be restricted to the marshy parts of Cambridge, Huntingdon, and Norfolk. It has been taken in many other places, but I believe that in all those instances it was not native to the place, but had been artificially introduced. In 1845 or 1846 I once saw a specimen in a field by the Cherwell, close to Oxford, and chased it for some time, but unsuccessfully. Whether or not this was an introduced specimen, I have no means of ascertaining.

The egg of this insect is light green in colour and oval in shape. It may seem rather superfluous to say that an egg is oval in shape, but we shall presently see that many eggs of Butterflies are anything but oval in shape. Just before the egg is hatched, its colour darkens until it is nearly black, which in fact is the colour of the young caterpillar. As soon as the larva is hatched, it eats the shell of the egg in which it has been developed, and after every change of skin it eats in like manner the garment which it has thrown off. The colour of this caterpillar is a beautiful leafy green, the interstices between the segments being velvet-black. Upon each of the twelve segments of the body there is a black bar, which in all the segments except the second is adorned with six orange spots. There are other markings, but these are the most characteristic.

This beautiful larva feeds on several plants, such as the hog's-wort, or cow-parsnip (*Heracleum sphondylium*), the marsh-parsley (*Peucedanum palustre*), and even on the leaves of the common carrot, when nothing better can be obtained. Larvæ of this splendid Butterfly have been successfully reared upon carrot leaves.

When the caterpillar is full fed, it quits its food-plant, crawls up the stem of a weed, and there assumes the pupal form, binding itself to the weed by a sort of belt. This belt may be almost called a cable, for it is very stout and strong, as well as elastic, and will sustain a considerable tension before it is snapped. There are many British Butterflies whose pupæ are thus girt to the object on which they undergo their transformation. Mr. Newman gathers these together in a group, which he terms *Succincti*, or Girted Chrysalids. All these pupæ have the head directed upwards.

THE next family, the Pieridæ, is distinguished by

the fact that the hind wings form a sort of receptacle in which the abdomen lies. The larvæ do not possess the nuchal horn, and are wider in the middle and narrower at the two ends. The insects which compose this family are the most familiar of our English Butterflies, and are popularly known as White Butterflies. There are, however, several White Butterflies that do not belong to this group, and several that belong to it which can scarcely be called white. However, the popular name is expressive, though not wholly accurate.

We will proceed at once to the best known of all our Butterflies, the LARGE WHITE BUTTERFLY (*Pieris brassicæ*), which is drawn in the accompanying woodcut. As to the colours of this insect, they are simply black and white, so that the illustration gives us a very good idea of the colour as well as of the form of the insect. The specimen represented is a male; the female being similarly coloured, but without the two large black spots on the fore wings. In its larval condition this is a most destructive creature, and does great damage to the plant from which it takes its name. It is, in fact, so destructive that those who keep kitchen gardens, or are interested in agriculture, should ruthlessly kill

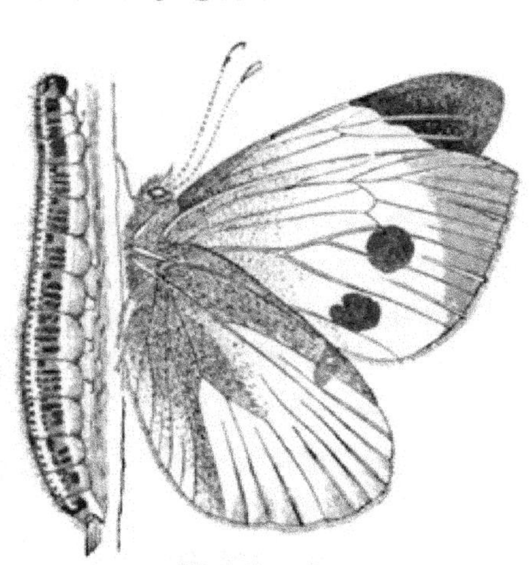

Pieris brassicæ.

every Cabbage Butterfly that comes across them. There is no cruelty in doing so, for no one has any scruple in killing the caterpillar, and it is surely better to kill one Butterfly than fifty or sixty caterpillars which it produces.

The reader may remember that, when treating of the Swallow-tail Butterfly, I stated that all insect eggs were not oval. The egg of this insect is an example of a considerable departure from the oval shape. These eggs are deposited in small clusters, varying from two or three to twelve, and they all stand upright on their bases, just like a number of little bottles, being fixed to the leaf by a gummy secretion. The caterpillars are hatched in about a fortnight, and grow very rapidly. When full-fed they ascend some convenient object, and change into the pupa, which is fixed by its tail, and prevented from falling by a stout silken belt passed loosely round its body. Its colour is grey-white, with a slight dash of blue, diversified by a number of little black spots. The point of the head is yellow, and so is a line along the ridge of the back.

There are two broods of this destructive insect, the first in May and the second in August, so that, if its numbers were not kept down by the ichneumon flies, we should scarcely have a cabbage in England.

THERE are several other Butterflies which go by the popular name of Whites.

There is, for example, the SMALL WHITE (*Pieris rapæ*), a very variable Butterfly, in which the male is nearly white, except a clouding at the tips of the upper wings, and a rather indistinct dark spot on the costal edge of the lower wings. The females are at once distinguished by having two black spots on the disc of the upper wings. The caterpillar is a pretty green larva, and is a very indiscriminate feeder. I have bred a long series of them from mignonette growing in a window-box. The Butterfly appears in the midst of summer.

The GREEN-VEINED WHITE (*Pieris Napi*) may be known by the peculiarity from which it takes its name. When inspected from below, the under surface of the lower wings is seen to be dull yellow, the nervures being edged with a greyish tint, which has the effect of green when contrasted with the yellow. The upper wings have much the same colouring, but not so strongly marked except at the tips. This caterpillar feeds on the watercress and one or two other plants.

CLOSELY allied to the Whites is the beautiful ORANGE-TIP (*Anthocharis cardamines*), so well known by the orange-tipped wings of the male, from which it derives its popular name. The female is without the orange hue, but both sexes have the under side of the lower wings beautifully mottled with green. The insect is as plentiful as it is beautiful, and may be captured in almost every meadow or lane in the early summer. It does not fly very fast, and generally keeps rather low, so that there is no difficulty in taking it. The caterpillar feeds on various cruciform plants, and its colour is opaque green.

IN the accompanying illustration may be seen a

drawing of a very well-known insect, the BRIMSTONE BUTTERFLY (*Gonepteryx rhamni*), the popular name being derived from the beautiful deep yellow of the male. The female is very much paler, as if the colour had been washed out of her, and in both sexes there is a little orange-red spot on each wing, the position and shape of which are indicated in the illustration. In this genus the wings are boldly angled at their tips, from which circumstance the name of *Gonepteryx*, or 'angle-winged,' has been given to it.

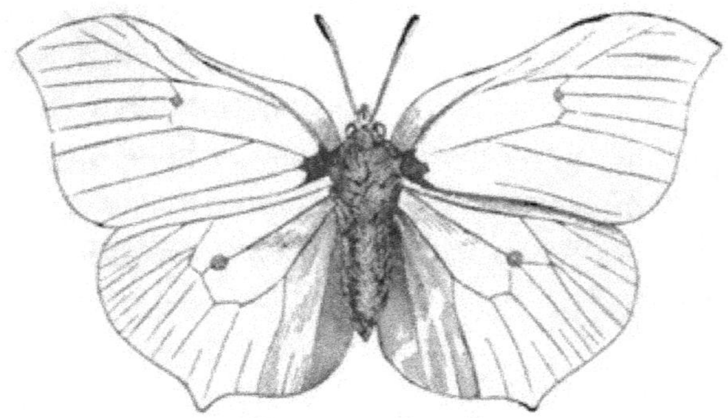

Gonepteryx rhamni.

This is one of our earliest Butterflies, sometimes making its appearance even in winter, should the sun happen to shine brightly. These early specimens are, in fact, the Butterflies which are developed in the autumn of the preceding year, and which had retired to some secluded spot wherein to hibernate. The warmth and light of a bright, sunshiny day awake them from their torpor, and tempt them into the open air.

The larva of this Butterfly feeds on the buckthorn (*Rhamnus*), from which it derives its specific name of *Rhamni*. The colour of the larva is green, but the surface of the body is thickly covered with tiny black projections or warts, each tipped with a slender white point.

BELOW is represented one of the prettiest of our Butterflies both in colour and form, the CLOUDED YELLOW (*Colias edusa*).

All those who have studied the Lepidoptera must have been struck with the marvellous variety and contrast of colour that can be produced by one or two hues. This insect is nothing more than black and orange, and yet is a singularly handsome one. The upper wings are warm orange, edged with a deep border of black, in which are a few pale orange spots in the female, the black band of the male being unspotted. There is also a bold black spot near the upper edge. The lower wings are coloured in much the same way, except that the orange is pale, and approaching to yellow. There is, however, a warm orange spot on the disc. Both pairs of wings are edged with a very warm border of orange, brighter and warmer in the upper than in the lower pair. Beneath, the colour is yellow, warming into orange on the disc of the upper wings, and the spot on the lower wings is brown, with a white centre.

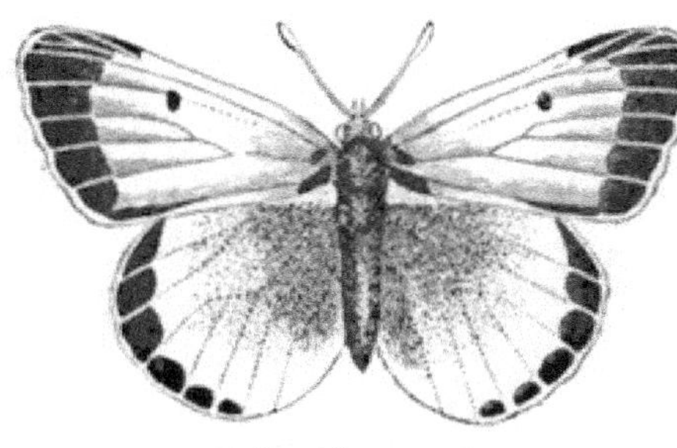

Colias edusa, male.

This handsome Butterfly is widely spread, and occasionally occurs in considerable numbers, especially in and near our Southern coasts. It is found in best condition at the end of summer, and through the autumn; but, as it is a hibernator, individuals are seen in the early spring, somewhat shabby and worn in appearance and lacking the brilliancy which distinguishes the newly-developed insect.

The larva is grass-green in colour, and is covered with a great number of tiny bristle-bearing warts. It feeds on the clover, and consequently clover fields are much haunted by the perfect insect. The eggs are long and narrow, and have been compared to nine-pins in shape. They are affixed to the leaf by one of the ends, so that they stand upright. The larvæ are hatched about Midsummer. The chrysalis, or pupa, is attached by its tail and a belt, like that of the Cabbage Butterfly, and does not seem to be very particular as to its position, generally being upright, but sometimes horizontal.

WE now come to the beautiful family of the Vanessidæ, which includes some of our handsomest and best-known Butterflies. Among other distinctions, in all these insects, the first pair of legs are very small, and not used in walking.

Our first example is the DARK GREEN FRITILLARY (*Argynnis aglaia*). The Butterflies belonging to this genus are popularly known as Fritillaries, and all of them have the under surface of the lower wings adorned with metallic spots and markings which look as if made of burnished silver. The generic name Argynnis, which is taken from the Greek, signifies 'brilliant' or 'shining,' and is given to the insects in consequence of this peculiarity. The colour of the present species is bright warm brown in the male, marked with a dark bronze-green in the female. The wings are profusely spotted with black on the upper surface, and on the under surface of the lower wings are a number of round bright silver spots, and near the margin of the wing is a row of seven semi-circular silver spots.

This insect is generally spread over England, and is plentiful in many parts of Kent, especially on the downs which lead towards the sea. Mr. Newman remarks that hilly, fern-covered ground is the best locality for this Butterfly. The larvæ feeds on the dog violet (*Viola canina*). Its colour is very dark shining grey, mottled with black. When full-fed, which takes place about the middle of July, it selects a suitable spot and there changes into a pupa, suspending itself by its tail, which is very strongly curved. It remains in the pupal state about eighteen days, and then emerges in its perfect form.

There are five other species of Argynnis, all of which are very similar in their colouring. As they are liable to variation, the beginner finds great difficulty in identifying them. This, however, can generally be done by means of the silver markings on the lower wings, which in some species form bold, clearly defined spots, like solid silver leaf, and in others take the shape of marks or streaks, just as if a brush had been dipped in silver powder and drawn over the wing. The larvæ of all the species are spiny, and feed upon the violets and their kin.

NEXT we take the typical genus Vanessa, of which we must examine several examples, as they are among the most conspicuous of our insects. In this genus the club of the antennæ is short and bold, the first pair of wings are more or less regulated, and the eyes are extremely hairy. If examined with the microscope, it will be seen that the hairs are planted at the angles of the hexagonal facets.

The handsomest and rarest of these Butterflies is called scientifically *Vanessa Antiopa*, and it is popularly called the WHITE-BORDERED, on account of the

Vanessa Antiopa.

broad white edging of the wings, or the CAMBERWELL BEAUTY, because in 1748 three specimens were taken near Camberwell, then a country village.

The colour of this magnificent insect is rich brown, shot with deep purple. The wings are edged with a broad grey-white band, just inside which is a row of blue spots.

ON the next page is a profile portrait of the

COMMA BUTTERFLY, scientifically termed *C. Album*, or the White C. These names are given to the insect because on the under surface of the lower wings there is a curved mark, very much like the letter C in shape, and of the purest white. There is not the least difficulty in identifying this insect, even without taking the trouble to inspect the under surface, for both pairs of wings are so deeply scalloped that there is no possibility of mistake. The upper surface of the wings is warm red-brown, mottled with black, both the ground hue and the markings being subject to considerable variation.

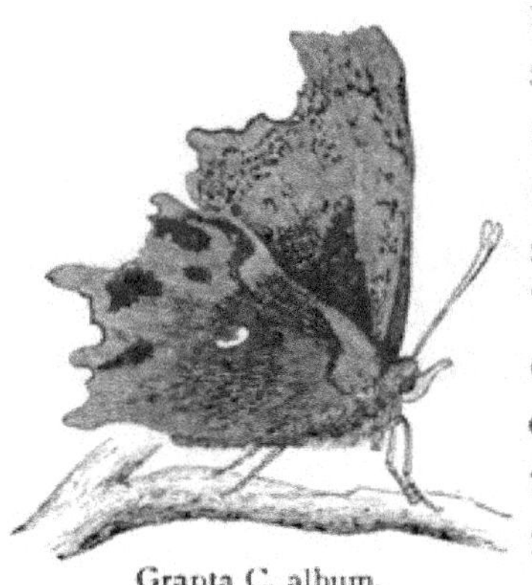

Grapta C. album.

A better known example of this family upon our list is the GREAT TORTOISESHELL (*Vanessa polychloros*).

This handsome insect is well named, as its rich mottlings of black and warm chestnut-brown bear no small resemblance to the colours of the tortoiseshell. In some parts of England this species is tolerably common, while in others it is never found.

It is common in Kent. Lieut.-Col. C. J. Cox, who has given great attention to this insect, told me an anecdote of the mode in which it deposits its eggs. He watched a female deposit an egg or two on a leaf, and wishing to rear the insect from the egg, he cut off the branch and removed it. The Butterfly, however, continued to fly up and down near the spot, and refused to leave it, evidently searching for the leaf on which she had deposited her eggs. The branch was then restored to its place in the tree, as nearly as could be done. The Butterfly at once saw and recognised it, proceeded to the same leaf, and deposited more eggs upon it.

Vanessa polychloros.

THE SMALL TORTOISESHELL, or COMMON TORTOISESHELL (*Vanessa urticæ*), is coloured much like the preceding insect, but the hues are rather brighter and the whole pattern of the wings defined more clearly, and looking more compressed. It is too familiar to need any detailed description. The caterpillar feeds on the common stinging-nettle, and may be found in great numbers upon it, sometimes being so numerous as to blacken it with the caterpillars clustering upon the leaves, which are drawn together with the silken threads spun by the larvæ. The pupa is suspended by the tail, and is very angular. Its colour is brown, mottled and spotted with black, and having several patches of a brilliant gold, as if burnished gold leaf had been laid upon it. This beautiful colouring has given to the pupa the name of Chrysalis, this being a Greek word signifying anything that is gilded. The golden hue is unfortunately very transient, and vanishes as soon as the Butterfly has emerged from the pupal envelope.

The splendid, and fortunately common, insect, the RED ADMIRAL (*Vanessa Atalanta*), comes next in order. This Butterfly can be at once recognised by the broad scarlet band near the upper wings and along the edge of the lower wings, a bold and conspicuous style of colouring possessed by no other British insect. The ground-colour of the wings is velvet-black, diversified with some large white spots on the tips of the upper pair of wings, and an oval blue spot on the anal angle of the lower pair. It is easy enough to describe the markings of the upper surface, but those of the lower surface are almost beyond description. Suffice it to say that the colour of the first pair is much like that of the upper surface only paler; while the under surface of the lower wings exhibits a most complicated mottling of brown, grey, blue, green, ochre, and black, arranged in a marvellously artistic manner, and forming a series of definite, but complicated patterns.

Vanessa Atalanta.

If this beautiful Butterfly were only rare, it would be the admiration of all collectors. As it is plentiful, it is only admired by those who value Nature, not for the sake of mere rarity, but for her own sake. Being one of the latter kind, I have a most enthusiastic admiration for the Red Admiral, and am never tired of examining it in the cabinet, or watching it as it flits at liberty in the open air, with the swiftly graceful flight that has earned for the insect the name of Atalanta.

The larva of this Butterfly is a nettle-feeder, and is quite common. The chrysalis, which may be found at the end of summer, is suspended by its tail from the roof of this habitation, and there hangs until the perfect insect makes its escape. This chrysalis, like that of the preceding insect, is adorned with brilliant golden patches on the sides, and its general colour is warm grey, mottled with black.

Ripe and fallen fruit has always a great attraction for this Butterfly. One of the most magnificent sights I ever saw was due to this predilection for fruit. A fine egg-plum tree had been entirely neglected, and its fruit permitted to ripen on the boughs and then fall to the ground. An army of insects took advantage of such an opportunity, and the tree and its vicinity swarmed with wasps, ants, and other lovers of sweets.

But the most striking point was the host of Atalanta Butterflies which surrounded the tree. They were approaching in every direction; the branches were crowded with them, and the fallen fruit upon the ground was so covered with them that neither fruit nor ground could be seen for the Butterflies, as they waved their black and scarlet wings gently up and down. So completely occupied were they with their rich banquet that they took no notice of me as I stood by them, and even permitted me to pick them up with my fingers. The sun was shining brightly on this wonderful assemblage, and brought out the grandly contrasted colours until they shone with tropical splendour. I never saw such a sight before, and am not likely to see such a one again.

WE will now pass to the pretty Butterfly called, on account of the variegated colouring of the under surface, the Painted Lady. Its scientific title is *Vanessa cardui.* The colour of the upper wings is deep black. There are five spots near the tip of a pure white, and the pale markings on the disc of the wings are warm chestnut. There is a slight white scalloping along the edge of the wing. The lower pair of wings are coloured in much the same way, but the only white portion is the scalloped edging. The bases of the wings are thickly covered with scales of a golden lustre.

In its larval state, the Painted Lady feeds on the thistle (*Carduus arvensis*), from which the insect derives its specific name of *Cardui.* It prefers the young and tender leaves of the plant, and draws their edges together, so as partially to enclose itself in the leaf. When it changes into a pupa, it suspends itself by the tail, and there remains until it emerges in its perfect form, somewhere about August.

In its habits, the Painted Lady much resembles the preceding insects, becoming developed in the autumn of one year, but not pairing until the spring of the next year. It is fond of flitting about in search of honey-bearing flowers, and especially frequents the teazle, on which flower the Red Admiral, the Great

Tortoiseshell, and the Peacock Butterfly often bear it company. I have taken all those insects plentifully on teazles in Bagley Wood, near Oxford.

As in the case with other Butterflies, the Painted Lady is wonderfully intermittent in its appearance, sometimes being absent or extremely scarce for several years, and then appearing in swarms for a year or two in succession.

WE must not complete our notice of this genus without brief mention of the beautiful PEACOCK BUTTERFLY (*Vanessa Io*), so conspicuous on account of the 'eyes' or circular marks on both pairs of wings. The under surface of the insect is brown-black, mottled in a most curious but almost indescribable manner. Whether the sombre colouring be intended for defence I cannot say, but there is no doubt that the insect often owes its life to the contrast between the upper and under surface. When a Peacock Butterfly is chased, it has a way of flying round a tree trunk, and settling on it, closing its wings at the same time, and bringing them together over its back. In this attitude it looks wonderfully like a dead leaf, and the change from a large, active, beautifully-coloured Butterfly, to a thin, black, shrivelled leaf, is so great and so rapid that scarcely any eye but that of an entomologist would detect the insect.

The larva is one of the nettle-feeders, and is mostly very common, in some places quite as plentiful as that of the Tortoiseshell, while in others it has to be searched for carefully before it can be found. The general colour of the caterpillar is black, and the

O 2

body is covered with a great number of long spines, which may, perhaps, defend it from the poisonous spikes of the plant on which it feeds. The chrysalis is green in colour, brighter when the change is first made, and darker when the future Butterfly is nearly developed.

IN the accompanying illustration is shown one of our handsomest Butterflies, the celebrated PURPLE EMPEROR, or EMPEROR OF MOROCCO (*Apatura Iris*). This genus differs from Vanessa in having the club

of the antennæ long instead of knobbed, and the club nearly straight. The eyes are without hairs, the hind wings scalloped, and the first pair of legs are not used for walking. Only one species inhabits England.

Its popular name of Purple Emperor is a very

appropriate one, at least as far as regards the male insect. The ground colour of the upper wings is brown-black in some lights, but in others is a rich shining purple, this effect being produced by the shape and arrangement of the scales, as can be seen by examining a specimen under the microscope. The light-coloured bands and spots are white, pure white in the male, and yellowish-white in the female. Beneath, the general colour is rust-red, blotched with black, grey, and here and there blue.

In former days the Emperor was supposed to restrict himself to the topmost branches of the oak-trees, and entomologists were accustomed to supply themselves with nets furnished with long handles made on the principle of the fishing-rod, the handle of the net being at least thirty feet in length.

Now-a-days entomologists who wish to catch the Purple Emperor do not trouble themselves to procure nets with preposterous handles—weapons with which I believe that a successful stroke can scarcely be made. They now look out for a secluded open space or glade in the wood, situated, if possible, on wet soil, lay on the ground a piece of bullock's liver, a dead rat or rabbit, or, in fact, any kind of animal substance, and go away again. For this insect has a strange liking for the most repulsive viands. There is nothing it likes better than the juices of putrid animal substances; and a dead dog or cat, which would drive away any human being who possessed nostrils and was not an entomologist, would attract the Purple Emperor to a rich banquet.

In two or three days, according to the weather,

they come back again, and steal quietly to the spot. Should the district be one that is favoured by this insect, very few of the baits will be without a Purple Emperor either settling on them or being at hand; and, like the Red Admiral, when engaged in taking food, the insect is so absorbed in its occupation that it can be taken without the least difficulty. If the locality be well selected, and the baits judiciously laid, it is very seldom that the entomologist will complete his round without having the opportunity of capturing this splendid Butterfly. So successful is this method of capturing the Purple Emperor that one entomologist succeeded in taking eighty specimens in nine days.

The caterpillar or larva is a very odd-looking creature, the most conspicuous points in which are the two horns with which the head is furnished. It feeds on the sallow, and, when partly grown, assumes so nearly the colour of the leaf that a sharp eye is needed to detect it. There are many markings and shades of colour in the caterpillar which need no notice, the general green hue and the horned head affording characteristics which cannot be mistaken. The perfect insect appears somewhere in July, the precise date depending much on the weather.

NEXT comes the family of the Satyridæ, in which the first pair of legs are not used for walking, the club of the antenna is bold and abrupt, and the wings are rounded. The larva has no spines, and the pupa is nearly smooth.

There are several well-known Butterflies belonging to this family, but our space will not permit of description. Such, for example, are the SPECKLED WOOD BUTTERFLY (*Satyrus Egeria*), the WALL BUTTERFLY (*Satyrus Megæra*), the GRAYLING (*Satyrus Semele*), the MEADOW-BROWN (*Satyrus Janira*), and the RINGLET (*Satyrus Hyperanthus*), the last-mentioned insect being remarkable for the fact that the upper surface of the wings is perfectly plain brown, while the under surface is warm brown adorned with sixteen spots, each spot having a white centre, then a broad black circle, and then an outer circle of white. It is an extremely variable Butterfly, both in the size and number of the spots; but ninety-nine specimens out of a hundred have the sixteen spots, three on each of the upper wings and five on each of the lower wings, the latter being arranged in two groups consisting of three and two spots. All these Butterflies are plentiful, and can be caught without difficulty in lanes or fields, their flight being sluggish, and never rising much above the ground.

WE now pass to the family of the Lycænidæ, in which are included those small, but very lovely Butterflies which are known by the popular name of Blues and Coppers, in allusion to the prevalent tints of their wings. The latter insects are seldom seen except by those who go to look for them, but the former are prevalent everywhere, fluttering low over wide downs, settling on wild flowers, or aiding in adorning our gardens with their beautifully variegated wings.

A GOOD example of the Blue Butterflies is depicted below. This is ths CLIFDEN BLUE (*Polyommatus Adonis*).

The colour of this pretty species is bright shining blue, with a delicate white line on the costal margin of the fore-wings, and a black line on the hind margin. The fringe of the wing is pure white, with a black spot at the end of each nervure. The female is brown on the upper surface, with a slight gloss of blue, and on the fore-wings is an indistinct blackish spot on the disc, and a number of small spots parallel with the hind margin. Both sexes have the underside of the wings brownish, with a number of little spots. This Butterfly is found on chalk downs and similar places, but seems to be restricted to those spots where there is a substratum of chalk.

Polyommatus Adonis.

PASSING of necessity by the rest of the Blues and the little Butterflies called by the name of Hairstreaks, we come to the family of Hesperidæ, popularly known by the name of Skippers, probably on account of their quick, uncertain flight. They are all small insects, coloured with brown, black, grey, and white, and very thickly and clumsily made, so that many persons unskilled in entomology take them for moths. They have the fore-legs fitted for walking, and when at rest they hold their wings only partially erect, and never pressed closely together, as is the case with the previously-mentioned Butterflies. They are plentiful in country lanes, and are often taken in the net when the stroke is made at other insects, their vague and almost jerking flight carrying them into it.

The pupæ of these insects resemble those of several moths in the manner in which they are protected during their helpless state of existence. When the larva is full-fed, it spins a cocoon among the leaves of its food-plant, and in that silken cell awaits its change into the perfect condition.

The species which is herewith represented is the Chequered Skipper (*Hesperia paniscus*). The colour of its wings is dark brown, upon which are a number of yellow spots, arranged as shown in the illustration. The antennæ are bright yellow beneath, and banded with black and yellow above. It is rather a local species, and is found chiefly in the Midland counties. It appears in the beginning of summer.

Hesperia paniscus.

CHAPTER II.

HETEROCERA, OR MOTHS.

The Nocturni, including the Sphinges and Bombyces of Linnæus.

IT has already been mentioned that there is no difficulty in distinguishing English butterflies from English Moths. Similarly, there is none in distinguishing English Moths fron English butterflies.

In the first place, the antennæ of these insects are not knobbed at the end, but pointed. Some of them have the ends of the antennæ much enlarged, as may be seen, for example, in those moths known by the popular title of Burnet. In all these, however, the club of the antenna is elongated, and the end is pointed. Moreover, whereas in the butterfly the shaft of the antenna is straight and simple, in many of the Moths it is curved, and in others is adorned with a feathering, sometimes on one side and often on both. For this reason the scientific name of HETEROCERA, or 'Varied horns,' has been given to these insects, just as Rhopalocera, or 'Clubbed horns' is given to the butterflies.

Then, the wings of a Moth at rest are not pressed together over the back, like those of the butterflies, but either lie flat on the body or along its sides. The body, moreover, has no waist like that of the butterfly. Keeping these few and obvious distinctions in his mind, the young entomologist need never hesitate to decide to which great group any lepidopterous insect belongs.

THE first family is that of the Sphingidæ or Sphinx-Moths. They derive their name, not from any peculiarity in the Moths themselves, but in their larvæ, some of which are thought to bear in their attitude a fanciful resemblance to the well-known Sphinx of Egypt. The Moths themselves are popularly and appropriately called Hawk-Moths, on account of the great swiftness of their flight, which very much resembles that of the hawk tribe. Their whole structure, indeed, shows that they are made for rapid flight, and, if we compare them with the swift-flying birds, especially the swallows and the humming-birds, we shall find that the outlines of Moths and birds are wonderfully similar. Their bodies are of moderate length, and pointed at the tail, and their wings are long, strong, stiff, narrow, and pointed. In fact, if the shadow of a humming-bird and of a Hawk-Moth were thrown side by side on the same surface, it would not be easy to tell which was the shadow of the bird and which of the insect.

With a very few exceptions, these Moths fly only at night or in the dusk of the evening, so that to watch their flight is not an easy matter. The best plan is, on some moonlight night, to take a stand near some honey-bearing flowers, to remain perfectly still, and watch quietly. Should there be Hawk-Moths in the neighbourhood, some of them are tolerably sure to come to the flowers, and to feed in their own peculiar manner, by poising themselves in mid-air on their rapidly quivering wings, and thrusting their sucking-tubes or trunks into the recesses of the flower. As these Moths will not fly by day, and as the partial darkness prevents their movements from being seen, it is better to look out for the well-known Humming-bird Hawk-Moth, which does fly by day, and is a very bold insect, allowing itself to be approached quite closely.

OUR first example of the Hawk-Moths is the magnificent DEATH'S HEAD MOTH (*Acherontia Atropos*), an insect seldom seen except by practical entomologists.

This splendid creature ranks among the very largest of our insects, inasmuch as the spread of its wings is very considerable, and the body is thick and heavily made. The upper surface of the fore-wings is warm-brown, with bands and mottlings of a darker hue, and a little white spot on the disc. The hind wings are yellow, with two black bands. The thorax is densely covered with a soft velvet-like down, feeling to the touch very much like the fur of the mole. The colour is a very deep black-brown, and in the middle is a yellow mark which bears the most startling resemblance to a skull and the two collar bones. The hair is so long that the shape of the skull can be altered by pressure. The body is yellow, with a longitudinal black stripe along the middle and six black transverse bands, each marking the edge of a segment. Beneath, the body and wings are yellow, with an indistinct dark band across the middle of each wing and a slightly darkened edging. The antennæ are very thick, covered with down, and furnished at the tips with a sharp hook.

The larva of this Moth is of enormous size, and is a very handsome creature, and exceedingly variable in tint, the general colour varying through different shades of yellow, green, and grey. The whole surface is covered with very tiny black dots, and on each side are seven diagonal blue or purple stripes, edged with white. Near the end of the tail is a curious horn-like appendage, curved downwards and then slightly recurved upwards near the point. The horn, contrary to the usual fashion of such horns, is very rough and covered with tubercles.

It feeds on various plants, of which the jessamine and the potato are the favourites, though it may be found on the snowberry, the tea-tree, and the deadly nightshade, this plant being allied to the potato. This caterpillar is invariably called a locust by the country people.

One of the most remarkable points in connection with this insect is its capability of producing sounds—a capability which is scarcely less striking than the skull depicted upon its thorax. If seized, or alarmed in any way, it produces—for I cannot say utters—a sharp squeaking sound, something like the cry of a mouse.

Although this sound is familiar to entomologists, no one has yet discovered its source. Some have thought that it is caused by the rubbing of the head against the thorax, some by the attrition of the antennæ and trunk, and some by the friction of the thorax against the abdomen. These theories are, however, neutralised by the fact that not only can the perfect insect produce the squeak, but that the caterpillar can do so, though it possesses neither trunk nor antennæ, and has no distinctive head, thorax, or abdomen.

Owing to the vast quantity of hair with which the body is covered it is of some consequence to secure specimens that are not damaged by being rubbed, as is generally the case with those that are captured by hand. The best mode of obtaining really perfect specimens is to rear them from the caterpillar. Labourers are not much afraid of the caterpillar, though they are of the perfect insect, and the easiest mode of obtaining both the larva and the pupa is to go to a potato-field in which the labourers are at work, and offer a small sum for uninjured 'locusts' and 'ground-grubs.'

The caterpillars of the Death's Head Moth being obtained and a continual supply of fresh potato-leaves ensured, they should be kept as much as possible in the dark. When they are full-fed they should be placed on light soft earth into which they will burrow, and undergo their transformations underground. It is as well to plant in the soil a few sticks up which the Moth can climb when it emerges, and to which it can cling while it dries its wings. Care must be taken to keep the earth moderately moist, placing damp but not wet moss upon it. Unless this precaution be taken, the outer skin of the pupa will become so hard that the insect will not be able to make its way out when it is fully developed. I have lost several Moths in this way, and have had one or two in a very maimed and imperfect condition, their wings being quite shrivelled, and scarcely one-sixth their proper size.

When the insect is killed the abdomen should be carefully severed from the body, and the whole of the contents removed by enlarging the little opening which will then be left at its base. The empty abdomen should then be stuffed with cotton wool, care being taken to make it full large in order to allow for shrinking, and when it is dry it may be joined to the thorax without leaving the least trace of the junction. It will be as well to pour a few drops of benzole into the abdomen and also into the thorax, as this precaution will keep off the mites and other creatures that work destruction among dried insects. All large-bodied Moths should be thus treated, and some of them can scarcely be preserved from the unsightly 'grease,' so hated by entomologists, without this useful substance.

For the purpose of rearing the Moth from the larva, the latter should be obtained about August, as it will then be nearly full-fed, and save a vast amount of trouble in procuring a supply of food. The pupæ themselves may be found under the soil somewhere about September.

OWING to our limited space, we can but casually glance at some other British Hawk-Moths. There is, for example, the EYED HAWK-MOTH (*Smerinthus ocellatus*), so conspicuous by the large eyelike spots in the middle of the lower wings and the beautiful pink brown of the upper wings. The larva of this Moth has a very rough skin, is pale-green in colour, speckled with white, and has seven diagonal stripes on each side of the body. The horn is blue.

Then there is the less conspicuous, but really beautiful, POPLAR HAWK-MOTH (*Smerinthus populi*), so common in the summer, clinging to the bark of trees, to rough posts, and other objects which somewhat resemble it in general colour. It may be known by the mottled brown of the upper wings, with a white spot in the middle, and the warm chestnut at the base of the lower wings. The caterpillar is rough, like that of the last-mentioned species, and is green sprinkled with yellow, and has seven diagonal yellow stripes on each side. The horn is yellow above and orange beneath. This caterpillar is plentiful, and can be beaten out of the boughs of the Lombardy poplar.

ON lime and elm trees may be found the larva of the LIME HAWK-MOTH (*Smerinthus tiliæ*). This is easily known by the very deep scalloping of the fore-wings and the prevalence of olive-green in its colouring. There is some variation in the arrangement of the markings, but the present species is the only one in which the deep olive-green is the leading colour, without any admixture of chestnut or pink. The attitude of the Moth at rest is a very curious one, the under wings being completely concealed beneath the upper pair, the scalloped edges of which, and their mottled surface, have the most astonishing resemblance to a pair of withered leaves.

The larva is pale green, and covered with very small tubercles, each being topped with yellow. Along the sides are seven diagonal stripes of yellow, which are mostly edged with pink. The horn is blue above and yellow beneath. In most parts of England this is a very plentiful insect, and can be either bred or captured without the least difficulty. I have bred considerable numbers of this insect, and have found no difficulty in rearing them—less difficulty, in fact, than I have experienced with any Hawk-Moth, except, perhaps, the Privet Moth, respecting which a few words must presently be said.

PASSING by the CONVOLVULUS HAWK-MOTH (*Sphinx convolvuli*), we may pause for a while upon the well-known PRIVET MOTH (*Sphinx ligustri*), so called in honour of the plant on which the larva feeds.

This fine Moth is really one of the commonest of British insects, although seldom seen on account of its nocturnal habits and the limited amount of the plant on which it mostly feeds. The Moth itself is a very handsome one, with a wide expanse of wings, very prettily coloured. The upper wings are very warm brown, mottled and clouded with dark brown; and the lower wings are pink, crossed by three nearly horizontal black bands. The body is pink, banded and striped with black. The caterpillar is a peculiarly handsome one. It is smooth and green, and has on each side seven diagonal stripes, the upper part of each stripe being violet, merging rapidly into white towards the under side.

These caterpillars feed on the common privet, and may almost always be found in profusion where that plant is present.

We now come to a very common and interesting insect, well known by the popular and appropriate name of HUMMING-BIRD MOTH (*Macroglossa stellatarum*).

The colours of this insect are anything but brilliant or conspicuous, and yet it is a very pretty Moth. The upper wings are brown, with a few slight black mottlings, and the lower wings are warm chrome yellow, with a narrow edging of black. Beneath, it is coloured much like the lower wings, but the hue is duller. The thorax and abdomen are of the same colour as the upper wings, but the latter has some black and white spots along the sides, which are covered with tufts of black and white hair, which are spread during flight. There is a tuft of black hair at the end of the abdomen.

Macroglossa stellatarum.

The caterpillar feeds chiefly on the Bedstraw (*Galium*), and, but for the characteristic horn at the end of the body, would scarcely be taken for the larva of a Hawk-Moth. Its colour is greenish brown, sometimes taking a pink tinge, and there are two lines along the sides, one pink and white, which reaches to the base of the horn, and the other dull brown, beneath the lighter line.

Sombre as is the colouring of this insect, I really do not know any Moth which is more interesting to the spectator. Fortunately, it flies by day, and, like the lovely bird whose flight it imitates, revels in the hottest sunshine. If, on a hot summer day, the observer will take his stand by a jessamine or other honey-bearing flower, and will quietly wait there, he will assuredly see a Humming-bird Moth before long, should the locality be one which is frequented by this insect. Suddenly, as he is watching a flower, his eyes see a kind of shadowy form flitting in front of the flowers, and his ears are greeted by the hum which accompanies the flight of the Moth. Let him but lift a hand, and the creature is gone—how, or where, it is impossible to say, so amazingly swift is the darting flight.

Still, though it be gone, it will come back again if no movement be made, and, in the same mysterious manner, the Moth is again hovering in front of the flowers. Presently, it selects one of them, and poising itself within an inch or so of the blossom, its body becomes visible, while its rapidly vibrating wings look like two grey patches of mist on the sides of the motionless body. Presently, a wonderfully long and slender tongue is thrust from the head, plunged deeply into the recesses of the flower, and, thus suspended in mid-air, the insect takes its sweet repast. It is a very remarkable fact that the Humming-birds themselves feed in precisely the same manner.

The enormously long proboscis or tongue, with which it extracts the liquid sweets from the flowers, has obtained for the genus to which this and a few other insects belong, the name of *Macroglossa,* or Long-tongue.

THE next family is that of the Sesiadæ, or Clear-wings, the members of which have a wonderful resemblance to certain bees, wasps, and flies, their wings being translucent, and their bodies being elongated and narrow, quite unlike those of the preceding insects. The antennæ have no feathering, and are very often tipped with a small tuft of hairs. The tongue is not nearly so long as in the preceding genus, and in most cases the end of the abdomen is tipped with a spreading brush of hair.

OUR first example of these remarkable Moths is the POPLAR HORNET CLEAR-WING (*Sesia apiformis*), which affords one of the best examples of imitation that I know. Only a few hours before writing this account, I was looking over some rather neglected drawers of insects, on the glass of which a slight layer of dust had been allowed to accumulate. I knew that a Hornet Clear-wing was among them, and yet the insect twice escaped observation, so strongly does it resemble the hornet both in colour and shape.

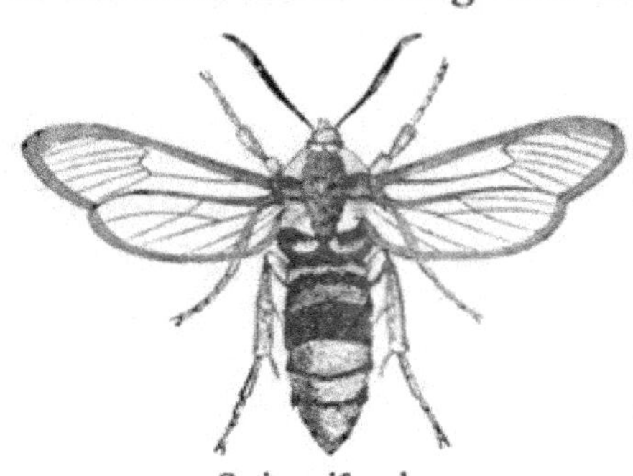

Sesia apiformis.

The upper wings of this insect are transparent, with a slight yellowish tinge, and a narrow, dark border. The head is yellow, and there is a yellow patch on either side of the brown thorax. The abdomen is yellow, with a broad dark-brown band near the middle, and a very narrow band of a similar colour nearer the base. The legs are orange-yellow. In fact, the colouring of this Moth is almost identical with that of the hornet, the peculiar rich, warm brown of the markings and the yellow of the ground colour being almost exactly identical in both insects.

The caterpillar of this Moth burrows into the wood of the poplar and aspen, and in its tunnel undergoes all its changes. When the larva is full-fed, it spins for itself a rather tough cocoon, made of small fragments of wood bound together with silk, and the Moth emerges about midsummer. The larva of this species passes two years in the tree before it changes to the pupal state.

The commonest insect of this genus is the Currant Clear-wing (*Sesia tipuliformis*), which may be found in the summer-time resting on the leaves of the currant. It bears a remarkable resemblance to a gnat, whence the name *tipuliformis*, i.e., formed like a gnat. There are two longitudinal yellow streaks on the thorax, and three bars of the same hue across the body.

The larva of this Moth lives inside the twigs and young branches of the currant, from which it bores out the pith, and often kills the branch. Indeed, whenever a bough of the currant begins to wither away without any perceptible cause, a larva of this Moth may generally be found within it.

THE family of the Zeuzeridæ is remarkable for the fact that the females are furnished with a long and hard ovipositor, by means of which they can introduce their eggs beneath the bark of the trees on which the caterpillars feed. The tongue and antennæ are short, and the larva is naked, with the exception of a few scattered hairs. There is a peculiarity in the chrysalis, which will presently be described.

A very characteristic example of this family is depicted below, the insect being popularly and appropriately called WOOD LEOPARD MOTH (*Zeuzera æsculi*). This is a very pretty moth, though the colours are simply white and black. The white,

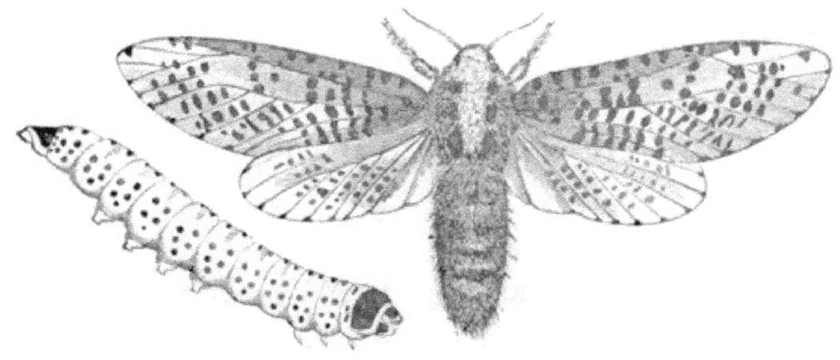

Zeuzera æsculi (female).

however, is partially translucent, and the black is in reality the very deepest blue-green. The figure represents the female. The male is coloured in a similar manner, but his antennæ are boldly curved, and adorned for the first half of their length with a deep double comb.

The caterpillar of the Wood Leopard very much resembles the perfect insect in markings, the green colour being white and the spots shining black. Behind the head is a large black plate. This caterpillar burrows into the limbs of many trees, especially fruit trees, but appears to do little if any harm to them. Indeed, Mr. Newman states that fruit trees which are pierced by this larva bear even more abundantly than those which are untouched by it.

The perfect insect appears in the middle of summer, and is a common insect, though it will seldom be found except by persons who know where and when to look for it. The female mostly remains near the tree in which she was bred, and may be found at night clinging to the trunk, where she can be detected by the aid of a lantern. The male is much bolder, and flies abroad in search of his mate. He is one of the many Moths that fly towards a light, and can often be taken by the simple process of putting a lamp near an open window. Many entomologists have made quite valuable collections of insects by this simple plan.

There is a peculiarity in the structure of the chrysalis of this and other Moths of the family. Each segment is furnished on its edges with a row of little hooks by means of which it can traverse its tunnel nearly as fast as it could while in the caterpillar state. As the pupæ of the Lepidoptera generally are quiescent, and few can do more than jerk the tail from side to side when irritated, it is rather startling to see a chrysalis wriggle itself up and down the burrow which it has made while in the caterpillar state.

WE now come to the GOAT MOTH (*Cossus ligniperda*), so-called on account of the very strong odour which is given out by the larva, and bears some resemblance to the powerful scent of the he-goat. A figure of this insect will be found on page 216.

The colour of this Moth is nothing more than

brown of different shades, arranged as shown in the illustration. As far as the perfect insect goes, there is but little interest about it, but in the larval condition it is peculiarly interesting, were it only for the fact that everything which can do harm contains within itself an element of interest. We will rapidly trace the life of the Goat Moth, from the deposition of the egg to the development of the perfect insect.

It has already been mentioned that the female

Cossus ligniperda.

Moths of this family possess a long ovipositor. With this instrument the female Goat Moth carefully lodges her eggs deeply in the crevices of the bark of some tree, and there leaves them. In process of time they are hatched, and, tiny as they are, soon are able to bore their way into the tree. They spend four years in the body of the tree, apparently never ceasing to eat, and boring their tunnels through and through the solid wood. These tunnels increase in size according to the development of the caterpillar, and, as a great number of larvæ generally inhabit one tree, it is no wonder that so many trees are killed by them.

The number of these destructive creatures that are found in one tree may be imagined from an experiment made by Mr. Douglas. He found an elm lying on the ground, having been blown down in consequence of the galleries of the Goat Moth caterpillar weakening the trunk so much that it could not endure the wind. He cut off a piece of the trunk two feet long, and twenty-three inches in diameter at one end and fifteen at the other. Wishing to place it in a vessel sixteen inches in diameter, he was obliged to cut it to the required size, and in so doing turned out no less than sixty-seven caterpillars, while the remainder of the block was equally well stocked with these larvæ.

The caterpillar is a very curious looking creature, and is well worthy of a somewhat detailed examination. The skin is very smooth and shining, and without hairs, except some stiff bristles which project from each segment. These bristles are not conspicuous to the eye, but are at once detected by the touch. The colour is pale mahogany, except the head and a hard plate on the second segment, which are black. The body is rather flattened, the head is wedge-shaped, and furnished with very powerful jaws. When full-grown the caterpillar exceeds three inches in length, and is as thick as a man's finger. The strength of a full-grown larva is enormous, and its powers of forcing its way out of the vessel in which it is confined must be seen to be appreciated. I have kept many of these larvæ, and never felt sure of them. They were continually escaping. Putting them into a wooden box was quite useless, as they ate their way through the side in a very short time. Putting them in a tin box was equally useless unless the cover were tied down, for they had a way of pushing at the lid round the edges, and so gradually opening it sufficiently to permit their escape. Even perforated zinc is not safe from them, for I have known my caterpillars to find out a place where the zinc has been cracked, fix their short and powerful jaws in the holes, and fairly twist down a flap through which they managed to force themselves.

When the larva is full-fed it forms a cocoon made of fragments of gnawed wood and silk. The cocoon is very tough, and will withstand much rough handling. These cocoons are oval, similar in shape at both ends, flexible, and yellow in colour, and are very strongly scented with the odour of the insect which made them. There is considerable variety in the size of the cocoons, and some are barely half as large as others. The small cocoons seem never to produce Moths, but are infested by an ichneumon-fly, which, fortunately for us, preys on the Goat Moth caterpillar. Not that the large cocoons are free from this parasite, for, as every one knows who has bred them, the large cocoons frequently disappoint the collector, and produce ichneumons instead of Moths. It is a curious fact that the ichneumon itself (*Lampronota setosa*) possesses an odour similar to that of the larva in which it lays its eggs.

After the larva has lain in its cocoon for some time, it discharges from its mouth a fluid which is contained in two large sacs within the body, and softens the silk so that it can be easily broken. It then throws off the caterpillar skin and becomes a chrysalis, which is at first white and soft, but afterwards hard and brown. The edges of the segments are furnished with little points directed backwards, and by alternately stretching and contracting the abdomen, the pupa forces itself along its larval tunnel until it comes to the end. Just before the final transformation the pupa renews its efforts, and fairly pushes itself through the thin shell of bark that has been allowed to remain by the larva.

It still continues to push its way on until it has forced itself through the opening, as far as the base of the abdomen. After a while the pupal skin splits, and the Moth emerges slowly, climbing up the bark of the tree, and there clinging while it shakes out its wings. The empty pupa skin remains at the entrance of the tunnel, and towards the middle or end of summer, according to the season, plenty of these empty shells may be found projecting from trees that are infested with the Goat Moth larva. The Moth itself can generally be captured upon the bark of the tree in which it has passed its pupal state.

The willow is the tree that is usually infested by this insect, and vast damage is often done by it. Other trees, however, are attacked by this destructive insect, especially the elm, and Mr. Newman is of opinion that those trees which are apparently killed by the Scolytus have received their death-blow from the Goat Moth, and have only been attacked by the Scolytus when dying.

In the illustration accompanying this description may be seen a common and very pretty Moth belonging to the family Zygænidæ. The insects belonging to this family may be distinguished by their very brilliant and boldly contrasted colours, their habit of flying by day, the stout and hairy caterpillar, and the cocoon fastened in an upright position against the stems of grasses. Only one genus inhabits England. The Moth which is given in the illustration is called scientifically *Zygæna* or *Anthrocera filipendulæ*, and is popularly known as the SIX-SPOTTED BURNET. There are several species of Burnet Moth, the greater number of which are so much alike that none but a practised entomologist can distinguish them, especially as the spots, from the number of which they derive their popular name, are almost precisely similar. This very beautiful insect has the upper wings of the deepest possible green, so deep indeed as to appear black unless the light be properly thrown upon it. The lower wings are rich crimson, edged with black. The peculiar form of the antennæ is so well shown in the illustration as to need no description.

Anthrocera filipendulæ.

PASSING of necessity over many Moths, we come to an insect which is both pretty and interesting. This is the CINNABAR MOTH (*Callimorpha Jacobeæ*), which is represented in the accompanying woodcut. It is an example of the family Euchelidæ, in which the antennæ are slender and without any fringe. The caterpillar spins a slight web, in which its hairs are scattered, and the pupa is small. The name Euchelidæ is formed from two Greek words, signifying Beautiful Caterpillar, and is given to the genus because the larvæ are all very beautifully coloured.

The colouring of the Cinnabar Moth is very bold, and is easily described. The upper wings are very deep olive-brown, looking almost black by the contrast with the brilliant crimson stripe near the costal margin, and the two crimson spots near the hind margin. The lower wings are wholly crimson, slightly paler than that on the upper wings, and are edged with a narrow border of olive-brown. It is remarkable that the upper and under surfaces of this insect are exactly alike, except that the under surface is slightly paler than the upper. The popular name of Cinnabar Moth is given to the insect on account of the cinnabar-crimson colour of its wings.

Callimorpha Jacobeæ.

ANOTHER family now comes before us — the Chelonidæ, popularly known as Tiger Moths. They have the antennæ more or less deeply fringed in the male, and the caterpillar is very hairy, coiling itself into a ring when alarmed. Before it changes into the pupal state, it spins a very loose web mixed profusely with its hairs.

The best known of these insects is the COMMON

TIGER MOTH (*Chelonia* or *Arctia caja*), so called on account of the bold cream and black markings of its upper wings. The lower wings are red, with some large black spots glossed with blue. The body is red, barred with black. This is a most variable insect, the number, size, shape, and tint differing in a most bewildering manner. But however variable it may be, there is never any possibility of mistaking it.

It is one of the commonest of British insects, and towards the end of summer the Moth is quite plentiful. It is wonderfully quick of foot, and, as it runs with closed wings among the herbage, has a most curious resemblance to a small mouse.

The larva is covered with long, brown, stiff hairs, and is popularly known by the name of the Woolly Bear. It feeds principally on the common dead nettle, but is not in the least particular as to its food, and, being very hardy, is an admirable subject for experiments in Moth-breeding.

PASSING by one or two well-known insects for which we have no space, such as the Ermines, the Brown-tail, the Golden-tail, the Gipsy, we come to that interesting insect, the common VAPOURER MOTH (*Orgyia antiqua*), the male of which is represented in the accompanying woodcut.

This is rather a pretty Moth, though the colours are anything but brilliant. The wings are warm chestnut, the upper pair having some waved transverse marks, as shown in the illustration, and a bold nearly semilunar white spot at the anal angle.

He is very common, and is one of the few Lepidoptera—except perhaps the Clothes-Moth, which is more plentiful than desired—that is abundant in London, and may be found even in the dingiest and smokiest portions, provided that trees or shrubs grow in it. He is one of the day-flying moths, and seems to revel in the blazing sun-beams, flitting about with rapid, and apparently uncertain wing, upon the hottest days of summer.

Orgyia antiqua, male.

There is, however, nothing uncertain about his flight, for he has a very definite object, namely, to seek a mate. Considering the kind of creature she is, and her peculiar habits, one is led to marvel, in the first place, how the active, prettily-coloured male Vapourer can find anything attractive in the female, who is about as plain—not to say, plebeian—an insect as can well be imagined. Indeed, a less attractive and more commonplace creature can hardly be seen. She has no wings to speak of, these organs being quite undeveloped and simply rudimentary, so that she could not fly one single inch. Her body is large, thick, soft, and covered with grey down, slightly darker at the edge of each segment.

This curious creature never wanders from the spot where she happens to have passed into the pupal state. Like the male, she has, when a full-fed caterpillar, spun a silken web, within which she has undergone her transformation.

The male has done the same, but when he has assumed the perfect form, he shakes out his pretty wings, takes to the air, and gaily sets out, like 'Cœlebs,' in search of a wife. She, on the other hand, never travels at all. Where she was reared, there she lives, there she is mated, there she provides a fresh brood, and there she dies, fulfilling the duties of her life within very narrow bounds. Her eggs are laid upon the silken web which she herself spun as a caterpillar, and from those eggs are hatched a brood of tiny larvæ, each of which is intended to follow in the track of its parents.

So plentiful are these egg-groups that, were it not for the presence of sundry little birds, which find much of their winter's nourishment in the eggs of various Lepidoptera, we should be soon overrun with Vapourer Moths, and our trees and hedges would suffer sadly. The female Moths themselves, being utterly unable to escape, and not seeming able even to crawl beyond the limits of the pupal web, also fall victims to the birds in no small number.

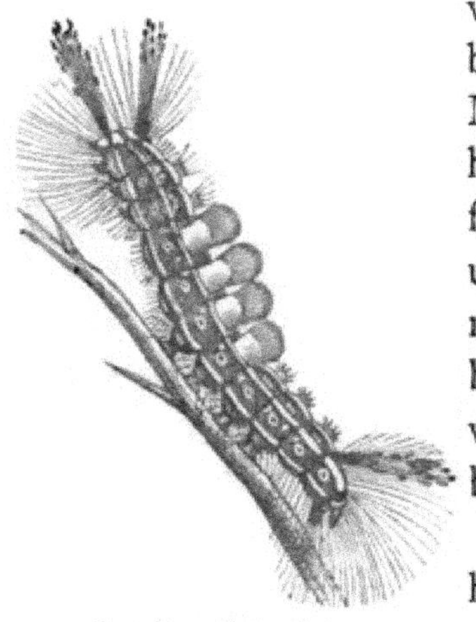
Orgyia antiqua, larva.

The caterpillar is shown herewith, and is a very pretty one. Its colours are exceedingly variable, but it is always furnished with a brush-like tuft of yellow hairs on the back of the fifth, sixth, seventh, and eighth segments, two long black tufts on the second segment, directed forwards, and a single similar tuft on the last segment but one, directed backwards. There is scarcely a tree or shrub on which this strange-looking caterpillar will not feed.

The well-known OAK-EGGAR MOTH (*Bombyx* or *Lasiocampa quercûs*) belongs to another family, the Bombycidæ, in which the caterpillars are mostly hairy, the pupa smooth, and the perfect insect large and stout-bodied, and coloured with various shades of brown or grey. The smooth pupa at once distinguishes this insect from the last.

Although the colours of this insect are not brilliant, the Moth is a very handsome one, the simple colouring of its wings being well contrasted. The male has the wings rich warm chestnut, and across each of them is drawn a slightly waved yellowish band. Rather towards the base of the disc there is a white spot very clearly marked. His antennæ are deeply and doubly feathered. The female is much larger, but not nearly so handsome, the colour being mostly yellow, with the band pale and undefined.

The chief interest of this moth lies in its preparatory stages. The caterpillar is a very fine one, and remarkable for its change in appearance when it bends its body. The ground colour of this larva is deep velvety-black, very thickly covered with rich brown hairs. When the caterpillar is straight it appears to be uniformly brown, but when it curves the body, the velvet-black appears between the segments and gives a very bold and effective appearance to the hitherto plain caterpillar. There are other marks, but these velvet rings are amply sufficient for identification.

It is very plentiful in some places, and though it is a very general feeder, eating almost every non-poisonous herb or leaf that may be given to it, the larva has fancies of its own and prefers one place to another, though apparently both localities are exactly alike in every respect.

When full-fed, the caterpillar spins a cocoon of wonderful toughness and strength. It is shaped very much like an egg (whence the popular name of Oak-eggar), and is brown and very close in texture. About the end of summer or beginning of autumn, the Moth breaks its way through the cocoon and appears in the perfect state. It mostly flies at night, but I have seen it on the wing at mid-day.

OUR last example of the Nocturni is the beautiful EMPEROR MOTH (*Saturnia carpini* or *pavonia-minor*).

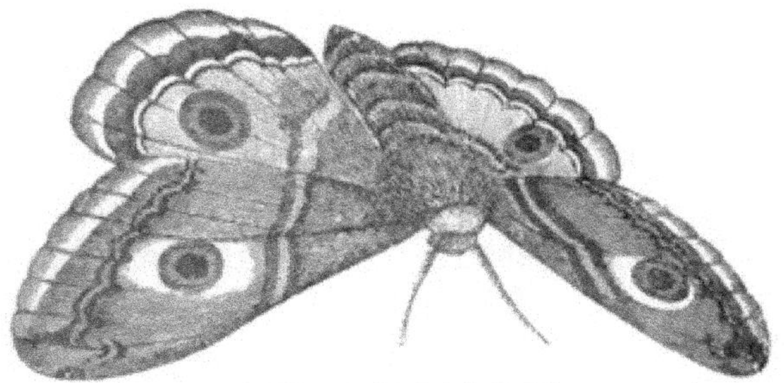

Saturnia carpini (female).

The general appearance of both sexes is very similar, except that the colours of the male are much more brilliant than those of the female. The colour and markings are so conspicuous that there is no necessity for very minute detail. The wings are pearly-grey, mottled and striped with brown, dark-grey, and chestnut. On each of the wings there is an eye-like spot, black in the middle, and surrounded with consecutive rings of warm buff and black, variegated with dark-crimson and violet.

The male has the upper wings of deeper and richer hues than those of his mate, and the under wings are warm ochreous orange, mottled and striped as in the female. The eye-like spots are similar in both sexes. The male is also distinguished by the antennæ, which are shorter than those of the female, and have a beautiful double feathering, widest in the middle, and decreasing towards the base and tip, so as to give the whole organ an outline much resembling that of the laurel leaf.

The caterpillar is quite as conspicuous as the perfect insect. It is beautiful leafy-green in colour, and the segments are marked so very distinctly that they look as if a number of threads had been tied tightly round the insect at the junctures of the segments. On each segment are a number of pink tubercles, each tubercle bearing a small brush of black bristles, and being surrounded with a ring of black. It feeds on a variety of plants, but I have found it more frequently on heath than on any other plant.

When it is full-fed, the larva spins a light-brown cocoon among its food, and the perfect insect appears in the middle of spring.

This cocoon is one of the most remarkable and interesting of insect habitations. Externally it is a simple brown, oval structure, more pointed at one end than the other, and having an outline much resembling that of a balloon. If it be carefully opened, and cut in two longitudinally, a most remarkable structure is seen. The smaller and pointed end is double, and within the outer case is a ring of short and stiff threads, looking much like bristles, their free ends directed towards the mouth of the cocoon, which is allowed to remain open. As these bristle-like threads follow the curve of the wall of the cocoon, it is evident that their ends must converge so as to close the opening against the entry of any insect foe, while they yield to the pressure of any creature within.

In consequence of this arrangement, the pupa remains securely shut up in its habitation, and, when the time comes for its assumption of the perfect state, the newly developed Moth creeps easily out of the cocoon, the guardian threads of which yield to its passage, and then close again, so that to all appearance the cocoon looks just as it did when it contained the chrysalis. As the caterpillar is a tolerably hardy one, there is no difficulty in obtaining the beautiful cocoons.

CHAPTER III.

GEOMETRÆ

THE large and important group of GEOMETRÆ, or LOOPERS, now come before us. These appropriate terms are applied to the Moths on account of the mode of progression adopted by the larva. The caterpillars are so constructed that they cannot walk after the usual fashion of such beings. The reader will remember that the caterpillars hitherto mentioned have a number of false legs, or claspers arranged on the under side of the body, in addition to the six true legs which are situated on that part of the body which will afterwards become the thorax of the perfect insect. Most caterpillars have five pairs of these claspers, but the Geometra larvæ have only two pairs, which are set closely together at the very end of the body. The caterpillar is therefore obliged to adopt a peculiar mode of progression.

When it wishes to move, it clings very firmly with its true legs, loosens the grasp of its claspers, and draws them close to the legs, so that its body is brought into an arch or loop. The claspers then fix themselves tightly to the object on which the caterpillar is moving, and the body is stretched out in order to find a fresh foothold for the legs. Thus, the caterpillar proceeds by bringing its body into the loop-like form and stretching it out for another hold. This may seem an awkward mode of progression, but it is nothing of the kind. There is even a sort of grace about the movement, and the caterpillar gets along at a wonderful pace, forming its successive loops with a rapidity that seems almost incredible.

The muscular strength of these caterpillars is wonderful. Most of us have seen acrobats fix their feet to an upright pole, or grasp it with their hands, and stretch out their bodies horizontally. This attitude requires great muscular powers very carefully applied, as those readers well know who have practically studied gymnastics. The leverage is so great that the strongest and most accomplished gymnast cannot maintain his position for any length of time, the attitude requiring the strongest possible strain on the muscles. Yet this attitude is not only easy to the Geometræ, but appears in some cases to be the chosen attitude of rest.

Several of these larvæ pass a large portion of their time stretched out at full length from the twig on which they are clinging. In this attitude they so exactly resemble twigs, that the sharpest eye can scarcely detect them, and even the most experienced entomologists are often deceived, taking veritable twigs for caterpillars, and caterpillars for twigs. None of the caterpillars are hairy, and their smooth bodies, often furnished with blunt spikes or humps, bear the most curious resemblance to the smooth-barked, bud-bearing twigs of the trees on which they live. Such caterpillars can be at once recognised as belonging to the Geometræ, and every entomologist knows that if he should find a looping caterpillar, and rear it, the result will certainly be a Geometra Moth of some kind.

In the perfect state it is not so easy to distinguish the Geometræ, though there is a certain and almost indescribable aspect about them that a practised entomologist rarely fails to detect, even though the species be new to him. We will now proceed to examine some of the most characteristic of these Moths in detail.

THE first family of the Geometræ is called Urapterydæ, or Wing-tail Moths, because in them the hinder wings are drawn out into long projections, popularly called 'tails.' In England we have but one insect belonging to this family, the beautiful, though pale-coloured, SWALLOW-TAILED MOTH (*Urapteryx sambucata*). The generic name is spelt in various ways, some writers wishing exactly to represent the Greek letters of which it is composed, and others following the conventional form which is generally in use. If the precisians are to be followed, the word ought to be spelled Ourapteryx.

There is no difficulty in recognising this Moth, the colour and shape being so decided. Both pairs of wings are delicate yellow, and the upper pair are crossed by two narrow brown stripes, which run from the upper to the lower margin. These stripes are very clear and well-defined, but besides there are a vast number of very tiny streaks of a similar colour, which look as if they had been drawn in water-colours with the very finest of brushes, and then damped so as to blur their edges. The hind wings have only one streak, which runs obliquely towards the anal angle, and, when the wings are spread, looks as if it were a continuation of the first stripe on the upper wings. The shape of the Moth almost exactly resembles that of the Brimstone Butterfly.

The perfect insect appears about July, and can be beaten out of bushes and hedges. Though the wings are large, they are thin and not very powerful, so that there is no difficulty in capturing the insect.

NEXT comes the family of the Ennomidæ, popularly called the Thorns, containing nearly thirty species, a typical example of which we will select for examination. In this family the hind wings are not tailed. Our first example is the BRIMSTONE MOTH (*Rumia cratægata*), which is shown beneath.

This very plentiful Moth is of a bright sulphur yellow, with a few irregular streaks, and several ruddy chestnut spots on the edge of the upper wings.

Rumia cratægata.

The caterpillar has three humps, and possesses four pairs of claspers instead of w o. The first and second pairs are, however, not used for progression. The larva feeds both on the blackthorn and whitethorn, and when full-fed spins a thick cocoon close to the ground, and sometimes on it. The Moth may be found throughout the summer, as may the caterpillar.

PASSING, of necessity, over many of the Geometræ, we come to that very familiar insect, the CURRANT MOTH (*Abraxas grossulariata*). In consequence of its boldly contrasted markings, it is sometimes called the MAGPIE MOTH.

This is invariably one of the first Moths of the young collector's cabinet, and its larva is perhaps the best known of the Geometræ.

The colour of the wings is white, with a yellow patch at the base of the upper wings, and a rather curved band of a similar colour rather beyond the middle. Upon both wings are a number of deep black spots and blotches, varying greatly in different specimens. Sometimes the black spots are so large that they unite with each other, and make the Moth look as if it were black and yellow. Sometimes the reverse takes place, and the insect is almost entirely white and yellow, with a few pale and uncertain markings of a darker colour, while in many specimens there is a decided preponderance either of the light or the dark portions of the wings. The antennæ of the female are thread-like, and those of the male very slightly, but decidedly, feathered.

This Moth is one of the partial day-fliers, and may be captured in any numbers in gardens where the gooseberry or black-currant is grown. The insect is a very bold one, and while it is engaged in depositing its eggs, may be picked up with the fingers without much difficulty.

The larva is coloured very much like the perfect insect. Being so common, I have been accustomed to watch it from childhood, and have much to say on the subject. But Mr. Newman has so completely made it his own that I can do no better than give his own spirited words.

'I have seen the females of this species busily engaged in oviposition, not only in the evening, but in the middle of a warm summer's day, depositing a single egg on a leaf of gooseberry or black-currant, and then flying off to another. I once watched ten females simultaneously occupied in this manner along a garden wall less than eighty yards in length.

'Like the eggs of most diurnal Lepidoptera, they remain but a short time before hatching. The young caterpillar feeds for two, three, or four weeks, rarely longer, and then spins together the edges of a gooseberry leaf, having first taken the precaution of making the leaf fast to its twig by numerous silken cables, which prevent the possibility of its falling when dehiscence takes place in the autumn. In the little cradle thus fabricated the infant caterpillar sleeps as securely as the sailor in his hammock. Snow-storms and wintry winds are matters of indifference to him, but no sooner have the gooseberry bushes begun to assume their livery of green in the spring, than instinct informs him that food is preparing to satisfy his appetite, so he cuts an opening in his pensile cradle, emerges, and begins to eat.

'The full-fed caterpillar commonly rests in a straight posture, lying parallel with the branch; but when annoyed, he elevates his back, and tucks in his head until it is brought into contact with the abdominal claspers. If the annoyance be continued, he drops from his food, hanging by a thread, and rarely falling to the ground ; but when this is the case, he is bent double, and remains a long time in that posture.'

In spite of the very conspicuous colouring of this caterpillar, it is not eaten by birds, seeming to be distasteful to them. It is also distasteful to toads. If one of these larvæ be placed before a toad, it will be snapped up as soon as it moves, but will at once be rejected, the toad moving off as if disgusted with a creature on which it hoped to feed. The colour of this larva is creamy white, spotted and striped with orange, and having a number of bold black spots and stripes, as seen in the illustration.

The pupa is smooth and black, with a slight tinge of red, banded with yellow, so that the caterpillar, the pupa, and the perfect insect have all the same colouring. The Moth appears in the middle of summer.

QUIET and simple-looking as is the WINTER MOTH (*Cheimatobia brumata*), which is represented in the accompanying illustration, there are few of our British insects which do more harm to the trees, especially the fruit-trees.

The colour of the upper wings is greyish-brown, more or less tinged with yellow, and marked with a few waved transverse bars of a darker tint. The lower wings are much the same colour, but almost without markings. Owing to its peculiar habits, this is one of our most familiar Moths. It appears in the cold months of November and December, and on a sunshiny day may be seen flitting along the hedges in perfect content, even though the ground be thickly covered with snow.

The female, though really the more important of the two, is seldom noticed. Her wings are mere rudiments, and she is unable to fly. She only appears at night, when she crawls up the stems of trees for the purpose of depositing her eggs upon them. When the little caterpillars are hatched, they make their way to the unopened buds and burrow into them, thus at the same time concealing themselves from sight, and doing all the harm of which so tiny a creature is capable. It is in search of these caterpillars that the small birds, more especially the bullfinch and chaffinch, pick off and devour the buds of fruit-trees. It is true that they do not restrict themselves to those buds which contain caterpillars, but that they act rather at random, picking off a bud first, and afterwards looking to see whether or not it contains a caterpillar. Still, the good that they do very much counterbalances the harm, and the little birds should be allowed to have their own way with the fruit-trees. The late Mr. Waterton would never allow a single little bird ever to be scared from his trees, much less killed, and I never saw anywhere better prospects of heavy crops.

Cheimatobia brumata.

Various plans have been tried to exterminate these mischievous caterpillars. Being silk-spinners, they lower themselves by their threads when alarmed, and, by taking advantage of this habit, the gardener can kill great numbers of the larvæ by simply tapping the boughs so as to frighten the caterpillars from their food. But 'prevention is better than cure,' and, on account of the structure and habits of the female, she can generally be prevented from depositing her eggs. Had she wings, nothing could be done; but as she is wingless, and is forced to climb up the trunks of trees before she can lay her eggs, it is mostly possible to prevent her from doing so. If the trunks of the trees be kept smeared with a sticky compound, renewed as soon as it begins to harden, vast numbers of the females can be interrupted in their march up the tree, and detained until they are slaughtered by the gardener.

Then, at night, the gardener should examine the trunks of all trees by the aid of a lantern, and he will be sure to find a number of female Winter Moths, each desirous of depositing her stock of two hundred eggs. I should fancy that birdlime might be useful. A mixture of Stockholm tar and cart-grease has been recommended; but if I had any standard fruit-trees, especially plums, greengages, or filberts, I should try the efficacy of bird-lime. In this part of the country, where cherry orchards abound, the fruit-growers paint the trunks of the trees with whitewash mixed with weak size. This process may or may not be efficacious, but there is no doubt that it is very unsightly, making the otherwise beautiful cherry-orchard an absolute eyesore.

When the caterpillar has become too large to be contained within the bud, it turns its attention to the young leaves, fixing its silken threads to their edges, and drawing two or three together, so as to form a sort of tent, in which it lives. It is not at all particular as to the tree on which it feeds, and, although it certainly prefers fruit-trees, may be found on almost every tree which is grown in England. The caterpillar is full-fed about May, and then descends to the ground, in which it burrows to a very little depth, and there changes to a pupa.

NEXT upon our list comes that very striking insect called appropriately the ARGENT AND SABLE (*Melanippe hastata*), which is shown beneath. As may be inferred from the name, the colours of this Moth are entirely black and white, mostly arranged as seen in the illustration, though there is some variation in different specimens. Even the body is black and white, and the very antennæ are black, with white rings.

The larva is rather variable, but is generally very dark brown, with a series of small black dots along each side, and another row of crescent-shaped white marks below the spiracles, each of which is surrounded with a white ring. It feeds on several trees and plants, such as the birch, and always lives in concealment, drawing together with silk the leaves of the plant on which it feeds, and living within this shelter. It is full-fed towards the middle of the autumn, and the perfect insect appears at the beginning of the following summer.

Melanippe hastata.

THE last of the Geometræ scarcely looks as if it belonged to that group. This is the common CHIM-

NEY-SWEEPER (*Tanagra chærophyllata*), which is represented herewith.

As may be inferred from its popular name, the colour of this Moth is sooty-black. The fringe is grey, except at the tip of the upper wings, where it is snowy-white. The larva is rather a pretty one, being dark green with a few lines of olive-green and light green. The spiracles are red. It feeds on the common earth-nut, or pig-nut (*Bunium flexuosum*), and is full-fed at the beginning of June, when it descends into the earth and changes to the pupal state. At the end of that month it assumes its perfect form, and in many localities appears in great numbers.

Tanagra chœrophyllata.

CHAPTER IV.

PSEUDO-BOMBYCES, DREPANULÆ, AND NOCTUÆ.

THE two first of these groups are gathered by Mr. Newman into one group, which are termed Cuspidates, because the tail of the larva mostly ends in a cusp or point. Some of the strangest caterpillars in the world belong to this group, and we have in sober England a number of Cuspidate larvæ which may rival the most wonderful productions of the tropics for beauty of colour and strangeness of form, the latter being in many cases actually grotesque.

The Pseudo-Bombyces are so called because the Moths look at first sight as if they belonged to the true Bombyces. The structure and habits of the caterpillar, however, show that these Moths are very rightly placed in a separate group.

The first family is the Dicranuridæ, so called on account of the structure of the larva. The name is formed from two Greek words, signifying Double-tailed, and is given to these insects because the tail of the larva is very deeply cleft, so as, in fact, to resemble two distinct tails.

The first of these insects is the common PUSS MOTH (*Dicranura vinula*).

This insect affords another example of the effect which can be got out of simple black and white. The upper wings are soft greyish-white and rounded at the tips, and have a peculiar softness in their general aspect. Most of the larger nervures are without scales, and show themselves conspicuously, but at the branches they are thickly covered with black scales. These wings are traversed by bold markings in black and dark grey, as seen in the illustration. The lower

Dicranura vinula.

wings are white at the base, deepening to blackish grey towards the margin, and having a few dark spots on the fringe. The large thorax is covered with long, soft down of a snowy-white colour, diversified with eight very black spots, so that it strongly reminds the observer of minever. The head is also white, and is held so much under the thorax that, when the creature is at rest, the head is quite invisible, and nothing can be seen but the ends of the antennæ, which are laid along either side of the thorax.

The caterpillar of this pretty Moth affords a singular example of grotesque form and beautiful colouring. The head is flat, and, when the creature is at rest, is drawn back into the second segment.

The fourth segment is produced into a large and pointed hump, and from the ninth segment the body tapers to the end. Here are developed two rough horn-like projections, from each of which can be protruded a horny pink filament, which seems to be employed as a weapon. It has been suggested that these appendages are used for the purpose of driving away ichneumon-flies when they settle on the body in the hope of depositing their eggs. Whether this theory be correct or not is undetermined, but the caterpillar certainly does protrude them when irritated. The larva has another weapon, if it may be so called. Below the head there is a transverse slit about the sixth of an inch in length. When the creature is alarmed or angered, from this aperture is ejected a fluid of an acrid character, which may probably have some injurious or deterrent effect upon the enemies of the Puss Moth larva.

The colour of this caterpillar is singularly beautiful—leaf-green on the sides and whitish above, with some stripes of purple brown. Between these two colours a white stripe runs from the side of the head to the tip of the hump, and then passes to the base of the double tail. The stripes are so arranged that when the larva is viewed from above, they appear something like the capital letter X. In some specimens, though not in all, there is a large purple patch on the eighth segment.

This larva feeds both on the willow and poplar, and, being very hardy, is easily reared throughout its changes. When full-fed, which takes place about the end of May, it leaves its food, crawls down the trunk of the tree, and creeps into some convenient crevice of the bark. In this refuge it forms a cocoon made of small chips of the bark fastened together with silk, and of wonderful strength. The cocoon, indeed, is mostly constructed of silk, the bark chips merely being added to it in order to make it agree in appearance with the trunk of the tree. Moisture does not soften this silken secretion, though air hardens it, and the consequence is that after exposure to the atmosphere, the cocoon becomes as hard as if made of horn, so that the inmate is safe from nearly all enemies ; while the exact similitude between the surface of the cocoon and the bark of the tree renders it almost incapable ot discovery.

The caterpillar is full-fed about midsummer, and, passing the winter in its chrysalis state, is developed into the Moth in the following June.

THE very common and really handsome Moth, the BUFF-TIP (*Pygæra bucephala*) is drawn in the ac-

Pygæra bucephala.

companying woodcut. The figure represents a male with its wings spread.

The upper wings of this Moth are beautifully

coloured with various shades of grey, crossed with bars and bands of different browns, mixed here and there with ochreous yellow, and taking a slight purplish gloss along the costal margin. At the tip of each wing is a large buff blotch, marked off from the rest of the wing by two dark-brown lines enclosing a grey line between them. The buff blotch has on it a few markings of deeper hue. The hind wings are paler greyish ochre, and the whole of the under surface is of the same colour, with the exception of a dark bar crossing the lower wings diagonally, and a dark hind margin to the upper wings. The thorax is large, covered with dense, gold-coloured down, and nearly conceals the head.

When at rest, this Moth presents a very curious aspect. The wings are pressed against the body which they cover, the two yellow spots at their tips exactly balancing at one end the yellow thorax at the other. The large thorax itself looks very much like a head, and on that account the specific name of *bucephala*, or 'bull-headed,' has been given to the insect. In consequence of the peculiar aspect of the quiescent attitude, the Buff-tip Moth often escapes observation, as most persons would mistake it for a piece of dried stick.

The caterpillar feeds upon the lime, the elm, the hazel, and one or two other trees, and often does considerable damage. It is semi-social in its habits, and, though plentiful, is not often seen until full-fed, in consequence of its peculiar idiosyncrasies. The eggs are laid in batches, sometimes as many as sixty in number, on the upper part of a leaf, and when hatched, the little caterpillars belonging to each brood remain together, and feed on the upper surface of the leaf. After their first change of skin, they break up into six or seven small companies, and each company remains together until the change into the pupal state is at hand. As they become larger they make their way to the topmost branches, where they work great havoc among the leaves, often leaving bough after bough completely denuded of foliage.

The colour of the caterpillar is yellow, covered with a number of short longitudinal black bars arranged in nine rows. The pupa is remarkable for the doubly forked apparatus at the end of the tail.

NEXT comes a family of Moths called Notodontidæ, or Tooth-backs, because the backs or inner margins of the upper wings are toothed, or have elevated portions along the inner margins, from which they derive the popular name of Prominents. As an example of these Moths we will take the IRON PROMINENT (*Notodonta dromedarius*), the male of which is shown beneath.

The colouring of this insect is very simple. The ground hue is brown, with a slight purplish tinge, upon which is a broad rust-red streak and two small pale bars, arranged as seen in the illustration. The outlines of all the markings are vague and indistinct, and there is considerable variation in different indivi-

Notodonta dromedarius.

duals. The lower wings are greyish brown, with a dark spot on the disc, and two pale and ill-defined bars.

The caterpillar is a very quaint and odd-looking creature. The head is comparatively large, and the second and third segments are so small as to form a sort of neck. From the fifth to the ninth segments the back is humped. The colour is rather pretty, being green more or less tinged with yellow, and marked with a very deep purple-brown. There are other markings, but the shape of the larva is so peculiar that minute detail is not needed for its identification.

This caterpillar may be found on the birch, where it remains until full-fed, an event which takes place somewhere about the end of September. It then descends the tree, and beneath it spins for itself a slight cocoon, which is generally screened from observation by having a fallen leaf fastened to its upper surface. In this exposed situation it changes into a pupa, and there lies until the following June, when it assumes the perfect form. The insect is, and yet is not, a common one. Those entomologists who have not yet learned to look behind the scenes of Nature's theatre reckon the Iron Prominent to be quite a rarity; while those who have been long accustomed to the practical study of insects and their ways, experience no great difficulty in obtaining either the moth, the pupa, or the caterpillar, and in consequence consider the Iron Prominent as rather a plentiful insect.

WE now come to one of the largest groups of British Moths, the NOCTUÆ, so called because, as a rule, they are exclusively night-fliers, and never, except by accident, appear in the day-time. In these Moths the body is almost always stout and thick, as is the thorax, the hairs of which often rise nearly erect, so as to form a sort of crest. Generally these Moths hide themselves by day, taking advantage of crevices in walls, the bark of trees, old posts, palings, and invariably selecting those which best harmonise with the colour of their closed wings. So close is often the resemblance between the colour of the insect and that of the object on which it rests, that even the most experienced and keenest entomologists often find themselves deceived, and have only detected the well-disguised insect when, by an accidental touch, they have forced it to take flight. Even those species which have their under wings adorned with beautiful colours, have almost invariably their upper wings plainly mottled with brown, grey, black, and dun, so that when they are at rest the splendid under wings are concealed, and their glories veiled by the sombrely tinted upper pair. Many, in fact, most of them, have both pairs of wings coloured in the simplest and least imposing manner, not even having any bold black, white, or brown markings on either pair of wings. Consequently, when a number of Noctuæ, which are of about the same size, are collected, it is a very difficult matter to refer them to their proper positions, and even the most skilful of entomologists is forced to refer to his books before he can, with any confidence, assign to some sixty or a hundred Noctuæ their exact names.

Most of the pupæ of the Noctuas undergo their changes beneath the surface of the ground, and the chrysalids that are found by the collector when 'digging for pupæ' generally belong to this group of insects.

OUR first example of the Noctuas is the pretty PEACH-BLOSSOM MOTH (*Thyatira batis*), which is represented in the accompanying illustration.

This very pretty Moth has received its popular name in consequence of the colouring of the wings. The upper pair are olive-brown, decorated with four large and conspicuous spots, the largest being at the base of the wing, and one smaller spot on the inner margin. These spots are lovely pink in the middle, surrounded with white, and each of them really does bear some resemblance to the petal of a peach-blossom. A few bars of rose-colour cross the brown thorax. The body is brown, and has a small crest on the back of the second, third, and fourth segments. The beautiful pink colour of the spots is very liable to fade, unless the insect be very carefully kept in the dark. The Moth is tolerably common.

Thyatira batis.

The larva of this insect is a very odd-looking creature. Its colour is warm chestnut-brown mottled with grey, and the surface has a velvety aspect. One peculiarity in this caterpillar is that it seems to make no use either of its true legs or of the claspers at the end of its body, but clings to its food plant by means of the claspers of the middle of the body. The largest hump is that of the third segment, and it is furnished at the top with a cleft projection. This curious larva can be found on the common bramble, and is in best condition about the end of August, or beginning of September, when it is full-fed, and about to 'spin up.' When it finally ceases to feed it spins a slight cocoon, which it fastens among the leaves, changes into the pupal state, and makes its appearance as a Moth in the ensuing summer. This pretty Moth used to be very plentiful about Oxford when I was collecting there. It belongs to the family Trifidæ.

WE must now pass to the only too common CABBAGE MOTH (*Mamestra brassicæ*).

That this Moth subserves some good purpose is evident from the fact of its existence, but what that purpose may be is not easy to discover. It may, perhaps, be useful in keeping down the too abundant vegetation in wild and uncultivated countries, and so may have done good service when this land was one vast hunting-ground, and our predecessors used flint instead of steel, and a wash of woad by way of dress. At all events, it is very much out of place so far as regards civilised society, and we could well spare it if it had been improved off the face of this country, in company with the wolf, the bear, and the beaver.

The caterpillar of this Moth is one of the most voracious herb-feeders in this country. It can eat almost any herb, but prefers those which belong to the cabbage tribe. As for those which are cultivated with solid masses of vegetation, such as the summer-cabbage and the broccoli, this larva is terribly destructive, burrowing through and through the very heart of the vegetable, and leaving behind it a track or gallery, filled with the watery juices of the plant and the ejecta of the caterpillars.

The colour of the upper wings of this Moth is dark greyish-brown, mottled variously with darker brown and grey. The lower wings are paler brown, with a smoky or blackish tint. The caterpillar is exceedingly variable in its colours, but is generally olive-brown above and yellow below, and on the back of each segment is a blackish triangular mark in which are two white dots. Sometimes the body is pale dusky-green above and below. When full-fed it descends to the earth, makes a shallow burrow in it, and changes to a smooth brown chrysalis. Both the Moth and caterpillar are plentiful through the summer, and during the autumn the ground may be nearly cleared of pupæ by judicious digging and hand-picking.

THE family of the Noctuidæ will be represented by two examples, the first of which is the TURNIP MOTH (*Agrotis segetum*), which is shown on page 251.

This is a small and inconspicuous Moth, but it does far more damage than many Moths of much larger size and more conspicuous colouring. The larva of this insect is to turnips what that of the last-mentioned insect is to the cabbage, and with this difference, that whereas the Cabbage caterpillar works above ground and may be detected by the eye, the Turnip caterpillar works for the most part below the surface of the earth, and the only evidence of its presence is the drooping state of the plant. When very young, it feeds upon the leaves of the turnip and many other plants, such, for example, as the carrot, or some flower, and in that stage may be removed by hand-picking; but, when it grows larger, it descends towards the earth, fixes upon the upper portion of the root, just where it joins the stem, and there gnaws a groove completely round the stem, the entire plant often dying from the injury. It grows with great rapidity, and, when about three-quarters grown, burrows into the earth and attacks the root itself, beginning near the bottom, burrowing deeply into it, and gnawing large hollows in it.

Agrotis segetum.

These caterpillars are only too familiar to agriculturists. They are nearly smooth, grey-striped more or less, and covered with little shining, round spots, from each of which proceeds a short bristle. As a rule, the gardener cannot mistake in killing every brown-looking caterpillar that he finds beneath the ground, for it is sure to be one of those beings that make havoc among the crops, and the greater because their ravages are carried on out of sight. It is principally in search of these destructive caterpillars that the rooks frequent turnip-fields. When the birds are seen busily digging with their powerful beaks, they are engaged in the search after the turnip caterpillar, and not trying to eat the turnip itself.

The whole history of this Moth is a very interesting

one, but our space is diminishing so rapidly that we must pass on to other insects.

THE LARGER YELLOW UNDERWING (*Tryphæna pronuba*), which is figured herewith, is by far the most abundant species of the genus to which it belongs.

Its upper wings are exceedingly variable in colouring, but are always of some shade of brown. There are several pale, narrow, waving bands drawn across the wing, and on the upper part of the disc is a large kidney-shaped black spot with a pale centre and a chestnut outline; there is also a small black spot near

Tryphæna pronuba.

the tip. The under wings are orange-yellow, but not so richly coloured as in the preceding insect, and parallel with their hind margin is a bold black stripe, broad above and narrowing below to a point. There is a very slight golden streak on its costal margin. Beneath, it is remarkable for a shining golden stripe that runs along the lower edge of the upper pair of wings, the gold changing in some lights to prismatic effects of green and blue.

The caterpillar is another of the nocturnal larvæ. It feeds upon the crown, stem, and heart of various garden plants, especially favouring lettuces when they are tied up to blanch. The colour of the larva is as variable as that of the perfect insect, but is generally some shade between olive green and brown, and on the body are a variety of brown and black streaks. It is a very general feeder, and there are very few garden plants or vegetables which escape its jaws. When full-fed, it forms a kind of rude oval cell, and therein undergoes its transformation into the Moth. Both the caterpillar and Moth are exceedingly common, and may be captured in any numbers.

PASSING over a considerable number of species, we come to an example of the Hadenidæ, the familiar ANGLE-SHADES (*Phlogophora meticulosa*), which is shown in the accompanying woodcut.

I have always felt a great predilection for this insect, because it is one of the first Moths that I ever reared. I had found a number of pupæ in the summer, and put them into a small box, which I covered with stout wire gauze, having in those days some hazy idea that a Moth could get through muslin or linen. I had also made up my mind that no Moths could emerge until the following year; and my astonishment was extreme on finding one morning a fine Angle-Shades Moth clinging to the wire gauze, and shaking out its newly-developed wings. Insigni-

Phlogophora meticulosa.

ficant as is such an incident in itself, it often forms a stand-point in life; and such was the case with this Moth, the development of which under my own eye inspired an interest in this branch of natural history that has never been and never will be forgotten.

The name of Angle-Shades is given to this Moth on account of the manner in which the wings are coloured. The upper wings are pale grey, tinged either with ochreous yellow or olive green. In the middle is a bold marking, shaped much like the letter V, and formed of several shades of brown. The other marks of the wing are also of brown, but not quite so dark. The edges of the hind margins of the upper wings are deeply scalloped. The lower wings are slightly scalloped, and are of a pale yellowish grey and crossed by two very slight waved bands of reddish brown. Towards the hind margins they are suffused with a slight pinkish tint. The thorax is covered with long hair, which in front stands out like the double ruff of Elizabeth's time. Then comes a wedge-shaped ridge in the middle of the thorax, and then two rather large tufts at the back. The colour of these tufts is soft umber brown, tipped with a darker and warmer brown.

The caterpillar varies in colour from bright green to dark olive green or olive brown, profusely sprinkled with whitish dots not very well defined. There are three greyish lines along the body, and the spiracles are white, surrounded with a black line. It feeds on various herbs and flowers, especially groundsel and primrose, and is full-fed about May, when it seeks the ground, and there spins a very slight cocoon. There are two broods of this pretty Moth—one towards the end of May, and the other at the end of autumn.

The specific name *meticulosa* signifies fearful or timorous, but I never could find out the reason for giving such a name to the insect. The Angle-Shades is not a whit more timorous than Moths in general, and though it has no distinctive boldness, it certainly has no distinctive timidity.

ACCOMPANYING this description will be found a representation of the Moth which is appropriately called the BURNISHED BRASS (*Plusia chrysitis*), in consequence of the metallic colouring of the wings. The specific name of *chrysitis*, or gilded, is given to it for the same reason. This insect belongs to another family of the Noctuæ, namely, the Quadrifidæ or Plusidæ.

The colour of the upper wings is bright golden green, which must be seen in a side light before its beauty can be properly distinguished. There is a large and nearly triangular blotch of brown on the middle of the wing, the base of the triangle resting on the costal margin, another patch of the same colour at the base, and a third on the inner margin, just below the large triangular patch. These two often coalesce, as is the case with a specimen now before me. The hind wings are greyish brown, and so is the body.

Plusia chrysitis.

The caterpillar is green, with a row of white dots under the spiracles, a white streak above them, and six white marks on the back of each segment. It assumes a curious attitude when at rest, the front of the body being bent upwards, so that the caterpillar only holds to its food-plant by its claspers. There are two broods of the Burnished Brass Moth in the year—one in the early summer, and the other in the middle of autumn. It feeds on several plants, such as the common white dead-nettle, and even the stinging-nettle.

To the same pretty genus belong several other well-known Moths, such as the SILVER Y (*Plusia gamma*), so easily recognised by the bright silver mark in the middle of the upper wings, closely resembling the English letter Y or the Greek letter gamma (γ). Then there is the BEAUTIFUL GOLDEN Y (*Plusia pulchrina*), the upper wings of which have a Y-like mark of burnished golden scales, and below it a round spot of the same colour. Another of these Moths is the GOLD SPANGLE (*Plusia bractea*), in which the upper wings have on the disc a moderately large and nearly square spot, which looks as if a patch of gold-leaf had been placed on the wing, and brilliantly burnished.

IF the reader will refer to the accompanying woodcut, he will see a portrait of the well-known HERALD MOTH (*Gonopteryx libatrix*), our only British representative of the family Gonopteridæ.

Even were not the colour of this insect so conspicuous, it could at once be identified by the shape of its upper wings, the hinder margins of which are deeply cut and scalloped, very much like those of the Comma Butterfly. The colour of the upper wings is soft brown-grey, with a downy surface, and slightly powdered with rust-red. On the middle of the wing is a broad dash of bright rust-red reaching as far as the base, and having a tiny, but conspicuous spot of pure white in its middle. Parallel with the hind margin a whitish-grey line runs across the wing and has a narrow, pale brown streak accompanying it. The front of the thorax is furnished with a ruff of long, soft down, of the same rust-red as that of the wing. The rest of the body and the lower wings are greyish-brown. The caterpillar is green, with a narrow grey streak along the sides. It feeds on the Sallow, and when full-fed spins a cocoon within two or three of the leaves, which it draws together with silk.

Gonopteryx libatrix.

The popular name of Herald is given to this Moth because it appears at the end of autumn, and is supposed to be the herald of the coming winter. Though feeding in the open air, it has a singular predilection for the habitations of man, and contrives to make its way into stables, outhouses, and even into houses that are inhabited.

JUST as one group of Moths is popularly termed the Yellow Underwings, so is another termed the Red Underwings, the ground colour of their lower wings being brilliant red.

One of these splendid insects is the RED UNDERWING (*Catocala nupta*), which is represented below. This is one of the largest and handsomest of the group, though its colours are not quite so brilliant as that of the Scarlet Underwing, an insect not so common as the present species.

The upper wings are grey with a slight yellowish tint, and profusely covered with waved bars and other marks of black, nearly every such mark being

Catocala nupta.

accompanied by a grey bar of similar shape. The under wings are red, diversified with two black bars, one, a very broad one, parallel to the hind margin, and another, a comparatively narrow and much curved bar, running across the middle of the wing. Beneath, the upper wings are white, crossed by three broad black bars, and the lower wings are similarly coloured, but warming into light red towards the inner margin, and having two bars across them.

The caterpillar is grey in colour, not unlike the hue of the upper wings of the perfect insect, and sometimes has two black waved stripes on the back. I never saw this caterpillar, but Mr. Newman's account of its habits is so admirable that I transfer it to these pages :—

'It feeds on the Crack Willow (*Salix fragilis*) and, when closely adherent to the bark, is almost impossible to detect. I have sometimes found it by passing my hand gently over the surface of the bark about a foot below the branches of a pollard willow, when its cold, soft feel at once betrayed it. It spins a network cocoon among the leaves, or in a crevice of the bark about Midsummer, and changes to a smooth chrysalis covered with a purple bloom.'

The perfect insect appears about August; and though it may be common, it is not often seen, owing to its mode of concealment. It carries into its perfect state one of its caterpillar habits, and has a way of settling on the trunks of willow trees and closing its wings. In this position the splendid red under wings are completely hidden by the sombrely tinted upper pair, which so exactly resemble the colour of the bark that, even when the Moth is pointed out, very few can distinguish it. I well remember the first time of discovering one of these beautiful Moths. I was going to bathe in the river Cherwell, near Oxford, a stream which is bordered with willows. I happened to place my hand on the trunk of one of the willows, when out bounced a grand Red Underwing, startling me as much as a novice in shooting is startled by his first pheasant. I afterwards found that the Moths were tolerably plentiful upon these trees.

The generic name Catocala is formed from the

Greek, and signifies something which is beautiful beneath. The name has been given to these insects because their chief beauty lies in the under wings, which are hidden beneath the upper pair when the Moth is at rest.

CHAPTER V.

DELTOIDES, PYRALIDES, AND CRAMBITES.

I VERY much regret that there should be no simpler words which can be substituted for those which head this chapter. There are, however, none whatever, so we must be content to use the terms which are adopted by the best entomologists of the time. Indeed, the only group of which it is even possible to form a simple English word which fully expresses the character of the insects, is the first, which literally signifies Delta-like, and may be freely translated as Delta-Moths, because when the insects are at rest, their wings assume a shape which bears some resemblance to the Greek letter Delta (Δ). All the above-mentioned insects are small, but the number of species is enormous, for they reckon among their ranks more species than all the groups which have heretofore been described. As is the case with the Noctuæ, the Moths of each group bear a great resemblance to each other, and much afflict the mind of the collector by their prevailing similitude.

ONE example of the Deltoides will be sufficient for our present purpose, and we will select the BANDED

SNOUT (*Hypena rostralis*), which is represented in the accompanying woodcut.

This is one of the Moths which have received the popular name of Snouts on account of the extremely elongated palpi, which project in front of the head so as to look very much like a proboscis. The antennæ of these insects are simple in the females and tufted in the males; their bodies are slender and furnished with a tuft on the first segment.

The present species has the upper wings of a yellowish-brown crossed with a dark, grey-edged band. It is a common Moth, and one of the earliest to appear in spring. The caterpillar is long and slender, hairy, and when full-fed spins a silken web among leaves and then changes into a long and slender pupa, having the head portion much elongated in order to contain the 'snout' or elongated palpi.

Hypena rostralis.

NEXT come the PYRALIDES, which some authors class with the preceding insects. They include, among other insects, the Meal-Moths, and the beautiful group of Pearl-Moths, so called because the surface of their wings has a peculiar sheen, much resembling that of mother-of-pearl.

WE will take as our first example of these Moths the too familiar TABBY MOTH (*Aglossa pinguinalis*).

The Moth is rather a pretty one. Its upper wings are yellowish-brown, with a very dark and nearly black patch at the base of each wing, and a broad stripe of the same colour running parallel with the hind margin, and much wider above than below.

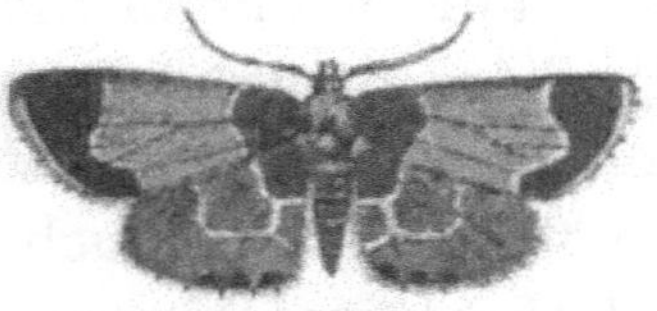
Aglossa pinguinalis.

A narrow white streak divides the dark from the lighter portions of the wing. The lower wings are of the same pale dun as the upper pair, but they are without the dark markings, and have only two jagged narrow streaks of white near them! When the wings are expanded, these marks look as if they were continuations of the corresponding marks of the upper wings.

This may almost take rank as one of the Clothes Moths, as in the larval state it feeds on old and greasy clothing. Grease, indeed, seems to be a necessity with this insect, which delights especially in old horse-rugs that are neglected by careless grooms. The specific name of *pinguinalis* signifies fatty or greasy, and is given to the species on account of the substances on which the larva feeds.

The caterpillar, which does the mischief, is a brown creature with a hard and horny skin, and having a head darker than the rest of the body. Like the ordinary Clothes Moths, it does not meddle with articles that are either in common use or that are carefully aired and looked after. But, should a groom throw a horse-rug into a corner, and let it lie there for several days, the larvæ of the Tabby Moth find their opportunity, and make sad havoc with the cloth.

When full-fed the larva makes a slight cocoon, and therein undergoes its transformations.

Now we come to a very extraordinary creature.

There is one family of Moths, comprising only four species, the larvæ of which are dwellers in the water, thus trespassing on the domains of other orders of insects. This is a group of Moths popularly called China Marks, because the general character of the surface of the wing and its markings has very much of a porcelain character about it. The typical species is *Hydrocampa stagnalis.* In all these Moths, the female is considerably larger than the male, and is rather variable in her colouring, so that the older entomologists have in several cases considered the sexes as forming distinct species. The male has no feathering to the antennæ, and the palpi are short, close together, and directed upwards.

The Moths are pretty little creatures, but the chief interest of the insect lies in the larva, which has a mode of existence that seems quite opposed to the whole character of the Lepidoptera. The caterpillars feed upon aquatic plants, and in some species are absolutely sub-aquatic themselves. It is evident that the respiratory apparatus of such larvæ cannot be formed like that of ordinary caterpillars, which breathe atmospheric air through spiracles and breathing tubes. Accordingly, these larvæ, like those of the caddis, the May-flies, and one or two beetles, such as the whirliwig, which has been described in the course of this work, are furnished with gill-like filaments along their sides, by means of which they extract the oxygen from the water just as fishes do.

This is a most wonderful fact, and almost without a parallel in entomology. There is one species of Ichneumon-fly, called *Agriotypus armatus*, which is so far aquatic in its character that it crawls down the sides of stones and water-plants to a considerable depth, evidently for the purpose of laying its eggs in some aquatic larva. It really seems to be fond of diving for its own sake, and if kept in an aquarium will submerge itself for a considerable time, the appearance of a hymenopterous insect beneath the surface of the water being most extraordinary, and always exciting the admiration and surprise of those who have any practical knowledge of insects. But, that a caterpillar should actually pass its life under the water is still more contrary to all preconceived opinions, and the idea of a water-caterpillar is not one whit more abnormal than that of a water-butterfly. The name Hydrocampa is formed from two Greek words literally signifying water-caterpillar, and is given to this genus of Moths in consequence of the aquatic life of the larva.

The Moths themselves are very common, and can be taken in plenty on the banks of ponds and any wet places where duck-weed, pond-weed, and water-lilies grow.

WE next come to the Pearl Moths, which are placed in the genus Botys.

In the genus Botys, the body is larger than the wings, and both pairs are marked in a similar manner.

The colour of our example of these Moths is pearly-white, on which are a number of dark markings arranged as shown in the illustration which will be found beneath. The popular name of this insect is the MOTHER OF PEARL, and its scientific name is *Botys urticalis.*

Although the general character of these marks is the same in all species, there is some variation in different specimens, both in their arrangement and depth of tint. In colour they are nearly black, but if viewed by a side light, a purplish metallic gloss is seen upon them, being best defined along the costal margin of the upper wings. Both surfaces are coloured in much the same manner, but on the under surface the marks are not so dark, and the purple gloss is more conspicuous, especially on the lower wings. When closed, the wings assume a heart-like shape, and usually look very round, as if a flat plate of thin mother-of-pearl had been cut into the shape of a heart, and carefully painted with dark spots. The thorax is bright golden-yellow, the abdomen is black, each segment being edged with yellow, and there is a tuft of yellow hairs at the end of the tail.

Botys urticalis.

The caterpillar is one of the numerous nettle-feeders. It is thicker in the middle than at the ends, and so thin-skinned that it has a semi-transparent appearance when viewed against the light. Its colour is whitish-grey on the back relieved by a central black line, and the sides are green. It draws together the leaves of the nettle with silken threads, and so feeds in concealment. There are ten species of this pretty genus.

THE next group of Moths which comes before us is called CRAMBITES, this name being apparently derived from a Greek word signifying a kind of caterpillar. Whether or not this is the case I cannot say, but the word has long been accepted by entomologists. The reader will notice that all Moths which belong to this group have their specific names ending in 'ellus' or 'ella,' according to the gender of the generic name.

WE can describe one example only of this group. namely, the PEARL-STREAK VENEER (*Crambus hamellus*), which is depicted below. In all the insects belonging to this genus, the labial palpi are very long and at first sight look very much like a beak, or short proboscis. The upper wings are long, narrow, and convoluted in repose, and the antennæ are thread-like.

This beautiful little Moth is rather gaily coloured. The ground hue of its upper wings is rich dark brown, and parallel with the costal margin and just below it is drawn a narrow streak of pearly white, from which its popular name is derived. The hind margins of the wings are yellow, that colour being separated from the brown by a dark, wavy bar. The lower wings are much lighter in colour than the upper pair, and not nearly so handsome, their colour being pale grey-brown, with an edging of yellow like that of the upper pair.

Crambus hamellus.

It is not a very common Moth, but can be taken in open places in woods by beating the bushes. It makes its appearance in the very midst of summer.

THE last of the Crambites which we can examine is an insect with which all bee-keepers would very gladly dispense, as it plays much the same part with the bee-comb that the Clothes Moth does with wool, fur, or feathers. This is the little insignificant-looking

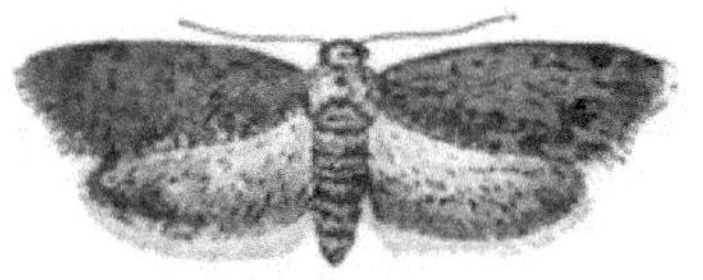
Galleria cerella.

Galleria, larva.

HONEY-COMB MOTH (*Galleria cerella*). The specific name of *cerella* (from the Latin word *cera*, wax) has been given to this insect in consequence of the wax-eating propensities of the larva.

A figure of the caterpillar is shown in the act of forcing its way through the honey-comb.

CHAPTER VI.

TORTRICES, TINEÆ, AND PTEROPHORI.

THE Tortrices or Twisters are so called because many, though not all of them, are in the habit of twisting or rolling up leaves while in the larval state.

There is little difficulty in knowing whether a Moth belongs to this group. In these, the body is comparatively short and slender, and the wings have a peculiar wave on their costal margin, so that when the insect is at rest with closed wings, the outline is curiously like that of a bell. We will take a few of the most conspicuous of these Moths.

AT their head come some Moths which scarcely seem to belong to the Tortrices. They are popularly called by the name of Silver Lines, because their green wings are crossed with some narrow lines of silvery-whiteness. As a rule, the Tortrices are all little Moths, but some of the Silver-Lines are exceptions to this rule, and are, indeed, the very giants of their race.

There are but three of these insects, which form the family of the Cymbidæ, a name which will be presently explained. The commonest of them is the

GREEN SILVER-LINES (*Halias fraxinana*). The upper wings of this Moth are beautiful leaf-green, across which are drawn three diagonal silvery lines, taking a pinkish hue near the inner margin. The head and thorax are of the same green hue as the wings. The lower wings and abdomen are pale yellow. The larva of this insect feeds on the oak, the ash, and one or two other trees, and the perfect insect appears in May. The middle of July is a good time for taking the larva, as it is then nearly full-fed. The colour of the caterpillar very much resembles that of the Moth's wings. This Moth measures about an inch and a half in the spread of its wings.

When the caterpillar is full-fed, it changes into a chrysalis, which is fastened to a leaf. The form of the chrysalis is most peculiar, and has been compared to that of a boat with the keel uppermost. The name of Cymbidæ, which has been given to this family, is taken from a Greek word signifying a boat, and alludes to the form of this pupa.

WE now proceed to the typical family, the Tortricidæ, of which we shall take two examples, the first of which is the very pretty, but very destructive PEA-GREEN MOTH or OAK-MOTH. The scientific name of this insect is *Tortrix viridana*. The appearance of this little Moth is very prepossessing, the upper wings being leaf-greèn, and the lower pair greyish-brown. When the wings are closed, the green is the only portion of the insect that is visible, so that the Moths may be thickly spread over a branch, and yet not one be distinguishable from the leaves. This insect is in some years very destructive among the oak trees. It may be found in abundance at the beginning of summer, in any place where oaks are numerous.

NEXT on our list comes the lovely, but destructive CODLIN MOTH (*Tortrix* or *Carpocapsa pomonana*).

This is a most exquisitely coloured insect, but a magnifying glass and a good light are required in order to bring out all its beauties. The upper wings are rich brown, banded at the base and tip with a darker and warmer brown. In the dark band at the tip of the wing is an oval mark of brilliant gold-coloured scales, having a very dark centre. In certain lights this dark centre takes a reddish hue, while a golden gloss pervades the whole of the wing. Even the outer wings, when viewed in a side light, shine as if made of the richest satin. By a proper adjustment of the light, a rather curious effect can be produced, the wings of one side glittering and shining in full splendour, while the corresponding wings of the other side are nothing but dull grey, brown, and black.

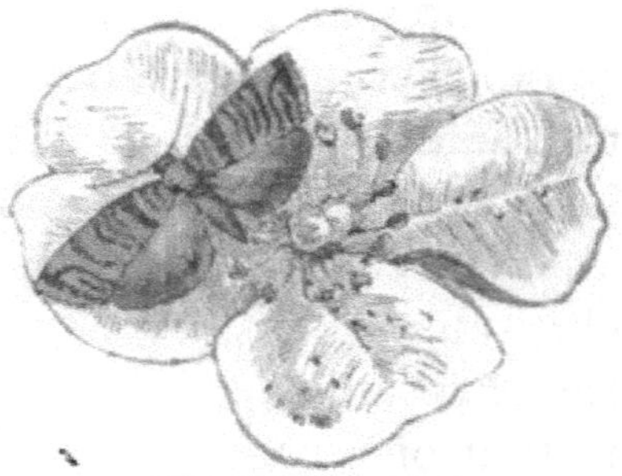

Tortrix pomonana.

Despite the destructive habits of this insect—perhaps in consequence of them—I have always cherished a kindly reminiscence of this Moth. When I was a child there was a remarkably fine codlin apple-tree in the garden, the fruit of which ripened early and was particularly juicy. As children, we were not allowed to gather the fruit at our discretion, but were permitted to take that which fell without any unfair means being used, such as beating or shaking the branches. Thanks to the Codlin Moth, a considerable number of apples always did fall annually, having ripened much before their time, as is the manner of fruit which will never come to perfection. Like most larvæ which never see the light until they are full-fed, the caterpillar of this insect is nearly colourless, with the exception of the head, which is brown-black, hard, and shining.

CHAPTER VII.

TINEÆ AND PTEROPHORI.

WE now come to a vast group of Moths, some of which are moderately large, while some are so very minute that they scarcely seem to be ranked among the Lepidoptera.

The name Tineæ is taken from a Latin word signifying a Clothes Moth, and has nothing to do with our word 'tiny,' however appropriate that may be in many cases. The number of these Moths is really unknown, for there is scarcely a year in which some new species of the Tineæ is not discovered and placed on the list. Indeed, so numerous are they that they have collectively been ranked under a separate name, viz. Micro-lepidoptera, and their study has become quite a distinct branch of entomology.

We will now proceed to examine a few examples of this group of Moths, and begin with the very beautiful and very mischievous insect called by the popular name of the LITTLE ERMINE (*Hyponomeuta padella*). With respect to the first of these names, I must mention that some writers on entomology omit the H and spell the word Yponomeuta. This practice, although it is followed by various writers, is utterly wrong, as it omits the aspirate in the Greek, the English representative of which is the letter H.

It is impossible to mistake the Little Ermine, whose long, narrow, satiny white upper wings, sprinkled with black dots, render it exceedingly conspicuous. The destruction wrought by this little insect is almost incredible, whole trees being stripped of their foliage, and, instead of bearing leaves, covered with the white webs and strong threads of the caterpillars. Even in the midst of London, in the densest, smokiest, dingiest part of Bermondsey, I have seen this Moth in full force. It was simply master of the situation. The little square yard which did duty for a garden was overrun with the caterpillars, which stretched their tough silken cables across the yard, across the windows, across the doorways, across the path, and, in fact, seemed to have calculated how they could most annoy the legitimate proprietors of the place.

Not content with taking possession of the tiny garden, they invaded the houses around, and every window that was opened was at once stormed by the caterpillars, which entered the rooms, crawled over the furniture, trailed their silken lines over everything in the room, and really made the inhabitants of the houses quite afraid to admit the little air that ever stirs in such localities. Yet, in the midst of all the smoke, the dust, the 'blacks,' and the other adjuncts of the neighbourhood, the little Moths fluttered about with wings as purely white as if they had never come within twenty miles of a chimney.

WE must not pass over without notice the lovely

Long-horn Moths, of which we have several examples in England. These Moths are remarkable for the extreme length and delicacy of their antennæ, these appendages being very much longer in the males than in the females. The best known of these is the DE GEEREAN (*Adela de Geerela*). This is a truly magnificent insect. Even to the naked eye its upper wings are singularly beautiful, but when it is examined by the microscope its splendour absolutely baffles description. Suffice it to say that the wings then appear to be covered with scale armour of burnished gold, every scale taking a rich purple hue in certain light. As the insect is turned under the microscope the edges become deeply purple, this hue being strongest and most conspicuous towards the tips. The fringe of the wings has also a tendency to purple.

The antennæ of this Moth are of enormous length. Just at the base they are rather thick, and have a very slight feathering. They then suddenly diminish, and are so long and so delicate that they almost look like the threads of a spider's web that have been casually attached to the creature's head. Indeed, I have often taken the Moth by watching for the flash of light reflected from the antennæ as they wave about in the air like threads of gossamer, while the insect is sitting quietly on a leaf.

THE next group of Tineæ are all leaf-miners in the larval state. The name of this group is the Lithocolletidæ. The object of this name I really cannot determine. It is derived from two Greek words, the former signifying a stone, and the latter to glue or cement. Taken collectively, the words may either signify a stone-cementer, or something that is inlaid or cemented with stone. Perhaps the colouring of the wings may have given some motion of a mosaic wall, which is made of small cubes of stone cemented together.

A FIGURE of one of the prettiest of these very pretty insects is given herewith, very much magnified.

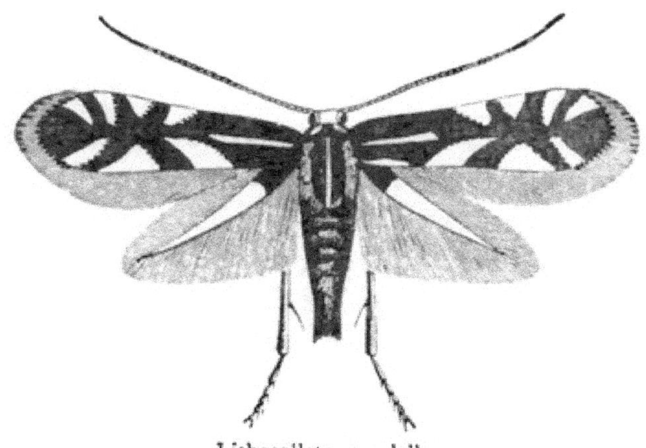

Lithocolletes corylella.

The scientific name of this insect is *Lithocolletes corylella*. No popular name has been given to it, so I shall call it the BROWN DOLLY, because the brown markings on the white wings bear, when viewed from base to tip, a certain resemblance to a rude wooden doll. It is really a very pretty insect. Viewed with the naked eye, it is so small that the shape of the markings is wholly invisible, and all that can be seen is a white surface profusely sprinkled with brown, or a brown surface spotted with white. But, when the magnifying glass is brought to bear upon it, the markings are seen to be very clearly defined. I have examined a considerable number of these beautiful little Moths, and in none of them was there any noticeable variation.

Although but few colours are employed in the decoration of the little Moth's plumage, it is a most beautiful insect, the rich brown and pure creamy white being contrasted in a wonderfully bold manner. There is a slight difference in the colour of the sexes, the upper wings of the male being soft creamy-white, while those of the female are cold grey-white. Upon their surface are drawn the rich brown markings shown in the illustration. The fringe is whitish-grey. The under wings are grey, fringed with a much lighter hue and darkening at the base. The sides of the head are yellow, and the face snowy-white ; the thorax is of the same colour as the upper wings, and the abdomen is grey, ended with a tuft of very pale yellow.

The larva is necessarily of very minute dimensions. When living it is about as long as the capital letter I, and, indeed, could not be much larger in consequence of the character of its home. Its colour is pale yellow, with a blackish-brown head, and a patch of deep orange upon the ninth segment. It burrows into the leaves of the hazel (*Corylus avellana*), from which it obtains its specific name *corylella*.

WE now come to some very curious and withal beautiful insects, though none of them possess the magnificent colouring which distinguishes the lovely little Moths which have just been described. They are scientifically known by the name of PTEROPHORI, or Feather-Bearers, and bear the popular name of

Plume Moths, on account of the structure of the wings. In those Lepidoptera which we have hitherto examined the wings are formed of a thin membrane stretched between certain strengthening nervures, or wing-rays, the principal of which radiate from the base of the wing. But in the Plume Moths there is no membrane, each nervure being furnished with long, hair-like plumes by which the insect is sustained in the air. In the under wings, the nervures are separate nearly from the base to the tip, but the upper wings are only divided from the middle.

THE commonest and one of the prettiest of the Plume Moths is popularly known as the LARGE WHITE PLUME, the SKELETON, or the PHANTOM, all names being perfectly appropriate. Its scientific name is *Pterophorus pentadactylus.*

This very beautiful though simply coloured insect has the wings pure snowy-white, and divided into separate plumes. In all cases the feathering of the plumes is much wider on the inner than the outer side of the nervure, very much like the structure of an ordinary bird's feather. If the wings be examined with the microscope, it will be seen that the long fringes which form the feathering are composed of the ordinary scales which cover the wings of the Lepidoptera, such scale being drawn out to a great length. Indeed, length of scale is one of the leading characteristics of these Moths, and the scales which clothe the base of the wings are also remarkable for their length. When viewed by a side light, these plumes have a satiny lustre, which quite disappears when they are viewed through a magnifying-glass, so as to render each of the delicate filaments visible.

THE Moth conceals itself during the day, making its appearance at dusk, when it flutters about like a snow-flake driven at random by the wind. It never makes a long flight, but if disturbed in one spot, just flits a yard or two and again settles on some leaf, where its white, outstretched, though not outspread wings render it very conspicuous when at rest. It never folds its wings to its body as do so many Moths, but remains with them stretched on either side to their very fullest extent, as if actually courting observation.

THE last of the Plume Moths, and indeed, the last Moth in our list of British Lepidoptera, is the beautiful little insect which is called by various names, only one of which is in any way correct. In some places it is known by the name of the THOUSAND PLUME, in others by that of the TWENTY PLUME, and in others the MANY-CLEFT PLUME, sometimes abbreviated into MANY PLUME. This last name is the only one which is correct in any way, and, after all, its correctness is only owing to its vagueness, which is almost a literal translation of its scientific name, *Alucita polydactyla*, or the Many-fingered Moth. In real fact the Moth has twenty-four plumes, which radiate from the body, so that, when the insect is at rest, its outline is almost semicircular.

It is but a little insect, the largest specimen measuring barely half an inch across the outspread wings. In its habits it is quite different to the Plume

Moths. They are always to be found in the open air, whereas the Many-Plume Moth is almost invariably taken in outhouses or similar buildings. I have frequently found it on the windows of my own rooms, its peculiar shape immediately betraying it. It can easily be taken by the plan called 'pill-boxing,' i.e. putting an empty pill-box over the Moth, slipping a piece of card or paper under it, and then putting on the lid of the box as the card is withdrawn. The box can then be put into the laurel bottle, or into a vessel in which a few drops of chloroform have been placed, and in a few minutes a perfect specimen will be at the collector's disposal. The colour of this little Moth is very pale brown, speckled with grey and dark brown, and taking an ochreous tint towards the base of the wings.

The larva of this Moth feeds on the buds of the honeysuckle, and is very common. About the end of July or the beginning of August the caterpillar is full-fed, and then spins for itself a cocoon, in which it undergoes its changes. This, I believe, is the only Plume Moth that makes a cocoon. In three or four weeks from the time of its change into the pupal state it is fully developed, and then makes its way to the nearest place of concealment, in which it may remain dormant throughout the winter.

INDEX.

LONDON: PRINTED BY
SPOTTISWOODE AND CO., NEW-STREET SQUARE
AND PARLIAMENT STREET

www.ingramcontent.com/pod-product-compliance
Lightning Source LLC
Chambersburg PA
CBHW021511210326
41599CB00012B/1214

* 9 7 8 3 3 3 7 1 3 9 8 0 3 *